Android Wearable Programming

Expand on your Android development capabilities by building applications for Android Wear

Steven F. Daniel

BIRMINGHAM - MUMBAI

Android Wearable Programming

First published: July 2015

Production reference: 1270715

Published by Packt Publishing Ltd.
Livery Place
35 Livery Street
Birmingham B3 2PB, UK.

ISBN 978-1-78528-015-3

www.packtpub.com

Cover image by Evelyn Lam (yeeyean@gmail.com)

Credits

Author
Steven F. Daniel

Reviewers
Marcus Gabilheri
Andreas Göransson
Dr. Jibo He
Qian He
Jason Salas

Commissioning Editor
Priya Singh

Acquisition Editor
Vivek Anantharaman

Content Development Editor
Amey Varangaonkar

Technical Editors
Utkarsha S. Kadam
Shiny Poojary

Copy Editor
Kausambhi Majumdar

Project Coordinator
Bijal Patel

Proofreader
Safis Editing

Indexer
Tejal Soni

Graphics
Jason Monteiro

Production Coordinator
Aparna Bhagat

Cover Work
Aparna Bhagat

About the Authors

Steven F. Daniel is the owner and founder of GENIESOFT STUDIOS
(http://www.geniesoftstudios.com/), a software development company based
in Melbourne, Victoria, that focuses primarily on developing games and business
applications for the iOS, Android, Mac OS X and Windows platforms. He is an
experienced software developer with more than 14 years of experience in developing
desktop and web-based applications for a number of companies, including: ANZ,
Department of Justice, BP Australia, and AXA Australia.

Steven is always interested in emerging technologies and is a member of the
SQL Server Special Interest Group (SQLSIG), Melbourne CocoaHeads, and
Java Community.

He was the cofounder and Chief Technology Officer (CTO) of SoftMpire Pty Ltd., a
company that focused primarily on developing business applications for the iOS and
Android platforms.

Steven is the author of various book titles, such as *Xcode 4 iOS Development Beginner's
Guide*, *iOS 5 Essentials*, *iPad Enterprise Application Development Blueprints*, and
Xcode 4 Cookbook, all by Packt Publishing. You can check out his blog at http://www.
geniesoftstudios.com/blog/ or follow him on Twitter at http://twitter.com/
GenieSoftStudio.

Acknowledgments

No book is the product of just the author — he just happens to be the one with his name on the cover. A number of people contributed to the success of this book and it would take more space than I have to thank each one individually.

I would personally like to thank two special people who have been an inspiration and who have provided me with so much support during the writing of this book: Vivek Anantharaman, my acquisition editor, who is the reason that this book exists, for being a wonderful guide throughout this whole process, and Amey Varangaonkar for his understanding and support, as well as his brilliant suggestive approaches during the chapter rewrites. Thank you for everything, guys.

Lastly, to my reviewers: thank you so much for your valuable suggestions and improvements, making this book what it is today. I am extremely grateful to each and every one of you.

Also, thanks to the entire Packt Publishing team for working so diligently to help bring out a high-quality product. Finally, a big thank you to the engineers at Google for creating the Android platform and providing developers with the tools to create fun and sophisticated applications.

Finally, I'd like to thank all of my friends for their support, understanding, and encouragement during the writing process. It is a privilege to know each and every one of you.

About the Reviewers

Marcus Gabilheri is a computer science student at Oklahoma State University. He was born in Brazil but lived in Spain for 11 years. He moved to the United States to be with his wife, Carissa Gabilheri, and decided to go back to school in the U.S. to follow his passion for programming. As a student at OSU, he has won the University's Mobile App Competition 2 years in a row. Marcus actively participates in the developer community as a Google Developer Group organizer and advocates Android development as well other technologies. He works as a mobile and web developer for Oklahoma State University and enjoys developing Android apps in his spare time. His latest achievement was in the Google Fit developer's challenge, where he was one of the 12 grand-prize winners of the challenge with his fitness app, FitHub.

> I would like to thank my wife, Carissa Gabilheri, for understanding and putting up with all the late nights and uncountable hours that I spend in front of my computer. I would also like to say special thanks to my mother, Adriana Andreo, and my grandmother, Maria Antonia Andreo, for raising me by themselves and to my parents-in-law, Randy and Donna Wilson, for accepting me into their family when I moved from Brazil to the U.S.

Andreas Göransson has been programming mobile phones since before smartphones became popular. He has previously written two books on Android development and has been involved in several open source projects. Beyond his interest in working on the family farm, he is very interested in emerging technologies, specifically, Internet of Things, wearable devices, and cloud-based services.

Dr. Jibo He is currently an assistant professor at Wichita State University. He graduated from Peking University in 2007 and the University of Illinois in 2012 with a research specialty in engineering psychology. He won the Star of Tomorrow Award from Microsoft and was voted the Most Valuable Graduate by the University of Illinois. He directs the Human Automation Interaction Lab at Wichita State University. His lab does research on user experience, mobile devices, driving safety, aviation psychology, and human computer interaction. The goal of his research is to understand the human cognitive processes and develop technologies to improve performance, increase user experience, and mitigate human error. He has experience in developing for Google Glass, Android, iPhone, and smartwatches.

Qian He is an enthusiastic digital gadget lover and experienced software engineer. He got his bachelor's degree in software engineering from Beijing Institute of Technology. After working at IBM and studying at University of Chinese Academy of Sciences, he decided to pursue a doctorate degree in the United States. Currently, he is studying computer science at Worcester Polytechnic Institute. His main research fields are ubiquitous computing and mobile health. Over the last few years, Qian has been interested in wearable devices and has built several famous fitness apps for Android / Android Wear / Pebble.

Jason Salas is a product manager, developer, sportscaster, author, and filmmaker who enjoys a perpetual summer on the island of Guam.

He runs the R&D group for Guam's largest media company, where he also co-anchors the nightly news.

Jason coauthored *Designing and Developing for Google Glass* (http://www.amazon.com/dp/1491946458/ref=cm_sw_su_dp), by O'Reilly Media, published an e-book about the trials and tribulations of a season in a semipro football league, and previously contributed to a book on Microsoft Hailstorm. He's also a member of the Football Writers Association of America.

You can find Jason at https://plus.google.com/+JasonSalas/posts.

www.PacktPub.com

Support files, eBooks, discount offers, and more

For support files and downloads related to your book, please visit www.PacktPub.com.

Did you know that Packt offers eBook versions of every book published, with PDF and ePub files available? You can upgrade to the eBook version at www.PacktPub.com and as a print book customer, you are entitled to a discount on the eBook copy. Get in touch with us at service@packtpub.com for more details.

At www.PacktPub.com, you can also read a collection of free technical articles, sign up for a range of free newsletters and receive exclusive discounts and offers on Packt books and eBooks.

https://www2.packtpub.com/books/subscription/packtlib

Do you need instant solutions to your IT questions? PacktLib is Packt's online digital book library. Here, you can search, access, and read Packt's entire library of books.

Why subscribe?

- Fully searchable across every book published by Packt
- Copy and paste, print, and bookmark content
- On demand and accessible via a web browser

Free access for Packt account holders

If you have an account with Packt at www.PacktPub.com, you can use this to access PacktLib today and view 9 entirely free books. Simply use your login credentials for immediate access.

To my favorite uncle, Benjamin Jacob Daniel, for always making me smile and inspiring me to work hard and achieve my dreams, I miss you a lot.

Chan Ban Guan, for the continued patience, encouragement, and support, and most of all for believing in me during the writing of this book.

To my family, for their love and support, and always believing in me throughout the writing of this book.

This book would not have been possible without everyone's love and understanding and I would like to thank you all from the bottom of my heart.

Table of Contents

Preface

Android Wear is becoming extremely popular, and offers a great opportunity for developers to learn how to build applications for the Android Wear platform, which is a special version of the core Android OS, and has been tailored for wearable computing devices such as smartwatches. These wearable devices come with a brand new user interface, which is the result of Google working with their customers to understand how they use their phones today and how they can be more in touch with their environment.

Android Wearable Programming provides a practical approach to developing and building Android apps using the Android Studio Integrated Development Environment. The new Android Studio IDE has introduced a specialized development environment that has been welcomed by the emerging Android community. This IDE is perfect to develop Android Wear apps because of the tight integration it has with the Wear development APIs, and also the streamlined build cycle with Gradle that helps to minimize a lot of manual configuration that the developer would need to do in other IDEs.

In this book, I have tried my best to keep the code simple and easy to understand by providing a step-by-step approach, with lots of screenshots at each step to make it easier to follow. You will soon be mastering the different aspects of Android Wear programming, as well as the technology and skills needed to create your own applications for the Android Wear platform.

Feel free to contact me at support@geniesoftstudios.com if you have any queries, or if you just want to drop by and say "Hello".

What this book covers

Chapter 1, Understanding Android Wearables and Building Your First Android Wear App, describes the background of the Android Wear platform architecture and shows you how to set up and configure the Android development environment, before finally looking at how to create a simple Android Wear app.

Chapter 2, Creating Notifications, introduces you to Android notifications, where you will learn how to create basic and custom notification messages. You will learn how to incorporate voice capabilities to read out the content of the notification, before learning how to group multiple notification messages using page-stacking.

Chapter 3, Creating, Debugging, and Packaging Wearable Apps, focuses on designing and creating custom watch faces to present information within the Android wearable watch area. You will learn how to effectively debug your app over Bluetooth, before finally learning how to package your wearable app so that it can be used within the handheld mobile device.

Chapter 4, Sending and Syncing Data, introduces you to the Data Layer API and the Message API frameworks, so that you can synchronize image data from the handheld device with the wearable, as well as use the Message API to communicate between the handheld and the wearable to send and receive messages.

Chapter 5, Working with Google Glass, explores how to build effective user interfaces for the Google Glass platform by creating user interfaces that display content that responds to voice input commands, before finally learning how we can access the Glass camera to take a snapshot and save the image to local storage.

Chapter 6, Designing and Customizing Interfaces for Android TV, provides you with the background and understanding of how to effectively present your app within the main user interface and how you can design your app by following the Android TV UI Patterns to help users get the content they want quickly. Also, you will learn how to create and use fragments that allow information to be presented within the Android TV interface to represent your content.

What you need for this book

For this book, you need a computer running a Windows, Mac OS, or Linux system. You will also need to have the Android Studio IDE and both Java and Java Runtime Environment installed on your system.

Who this book is for

This book is intended for developers who have a working experience of the application development principles for the Android platform and wish to expand their Android capabilities by developing applications for Android wearables using the key features of Android Studio. It's assumed that you are familiar with object-oriented programming and the Java programming language.

Conventions

In this book, you will find a number of styles of text that distinguish between different kinds of information. Here are some examples of these styles, and an explanation of their meaning.

Code words in text are shown as follows: "We can include other contexts through the use of the include directive."

A block of code is set as follows:

```
public class MainActivity extends ActionBarActivity {

    // Set up our Notification message ID
    int NOTIFICATION_ID = 001;
```

When we wish to draw your attention to a particular part of a code block, the relevant lines or items are set in bold:

```
dependencies {
  compile fileTree(dir: 'libs', include: ['*.jar'])
  compile 'com.android.support:appcompat-v7:21.0.3'
  compile 'com.android.support:support-v4:20.0.+'
}
```

Any command-line input or output is written as follows:

```
$ ./adb forward tcp:4444 localabstract:/adb-hub
$ ./adb connect localhost:4444
```

New terms and **important words** are shown in bold. Words that you see on the screen, in menus or dialog boxes for example, appear in the text like this: "Next, click on the **Install packages** button as shown in the preceding screenshot."

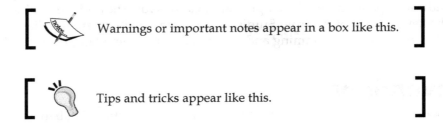

Warnings or important notes appear in a box like this.

Tips and tricks appear like this.

Reader feedback

Feedback from our readers is always welcome. Let us know what you think about this book—what you liked or disliked. Reader feedback is important for us as it helps us develop titles that you will really get the most out of.

To send us general feedback, simply e-mail feedback@packtpub.com, and mention the book's title in the subject of your message.

If there is a topic that you have expertise in and you are interested in either writing or contributing to a book, see our author guide at www.packtpub.com/authors.

Customer support

Now that you are the proud owner of a Packt book, we have a number of things to help you to get the most from your purchase.

Downloading the example code

You can download the example code files from your account at http://www.packtpub.com for all the Packt Publishing books you have purchased. If you purchased this book elsewhere, you can visit http://www.packtpub.com/support and register to have the files e-mailed directly to you.

Errata

Although we have taken every care to ensure the accuracy of our content, mistakes do happen. If you find a mistake in one of our books—maybe a mistake in the text or the code—we would be grateful if you could report this to us. By doing so, you can save other readers from frustration and help us improve subsequent versions of this book. If you find any errata, please report them by visiting http://www.packtpub.com/submit-errata, selecting your book, clicking on the **Errata Submission Form** link, and entering the details of your errata. Once your errata are verified, your submission will be accepted and the errata will be uploaded to our website or added to any list of existing errata under the Errata section of that title.

To view the previously submitted errata, go to https://www.packtpub.com/books/content/support and enter the name of the book in the search field. The required information will appear under the **Errata** section.

Piracy

Piracy of copyrighted material on the Internet is an ongoing problem across all media. At Packt, we take the protection of our copyright and licenses very seriously. If you come across any illegal copies of our works in any form on the Internet, please provide us with the location address or website name immediately so that we can pursue a remedy.

Please contact us at copyright@packtpub.com with a link to the suspected pirated material.

We appreciate your help in protecting our authors and our ability to bring you valuable content.

Questions

If you have a problem with any aspect of this book, you can contact us at questions@packtpub.com, and we will do our best to address the problem.

1
Understanding Android Wearables and Building Your First Android Wear App

When Google announced Android Wear at their Google I/O conference back in March 2014, developers were excited and started embracing this technology to see what types of applications they could create to communicate between the Android handheld device and Android wearable, while making our day-to-day lives a lot easier.

Android wearables bring a personal touch by allowing consumers to interact with their devices on a different level, and are aimed at reducing people's interaction with their mobile phones. This could include receiving a simple notification message reminding you to pick up something on your way home from work, or that you have an upcoming appointment. There is even an ability to receive messages from your favorite social networking application, for example, Facebook.

This chapter provides you with a theoretical background of Android, and how to develop applications for the Android Wear platform from Google. This platform allows your Android wearable device to communicate with your phone wirelessly over Bluetooth, and many manufactures like Samsung and LG have embraced this technology and created wearable devices, such as the Samsung Gear Live and the LG G Watch R smartwatches.

In later chapters, we will be working with some of these APIs and seeing how we can incorporate these in our applications to communicate between our Android phone and Android Wear devices.

This chapter includes the following topics:

- Introducing Android wearables
- Understanding the Android Wear architecture
- Building a simple Android wearable application

Introducing Android wearables

Android Wear is a special version of the core Android OS that has been tailored for wearable computing devices such as smartwatches. These wearable devices come with a brand new user interface, which is a result of Google working with their customers to understand how they use their phones today and can be more in touch with their environment.

Android Wear provides consumers with a more personal interaction with their devices. These tiny supercomputers can show you information and suggestions when you need them. Given the wide variety of Android applications currently on the market, you'll receive the latest posts and updates from your favorite social apps and notifications from shopping apps.

Android wearables are great for fitness fanatics too. They allow you to better monitor your health and fitness by showing your fitness summary in terms of real-time speed, distance, and time information right on your wrist for your run, cycle, or walk.

Android Wear also lets you access and control other devices from your wrist by simply saying OK Google to fire up a music playlist on your phone or cast your favorite movie directly onto your TV. You can also receive instant messages from your favorite social networking app, for example, Facebook. With Android wearables, there's a lot of possibilities, and developers are jumping right in and creating some stunning apps already.

An example of an Android wearable device can be seen in the following screenshot:

Understanding the Android Wear architecture

Android Wear works by communicating wirelessly over Bluetooth between the wearable and a handheld device (typically a smartphone) running Android 4.3 or higher. When the handheld device has been paired with the wearable device, the operating system begins sending a series of notification messages automatically to the watch, along with any wearable-specific rich notification parameters, such as voice input for actions and any specific pieces that provide additional information.

When a connection has been established between the Android device and the wearable, over Google Play service, notification messages can be then exchanged between the handheld device and the wearable to trigger appropriate actions on each device.

The architecture of any typical wearable application has been set out by Google in their design guideline documents that focus primarily on the new Material design theme for Android 5.0 applications. This design document provides the developer with a comprehensive framework to create visual, motion, and interaction design across each of the various Android platforms and devices.

Since Android Wear runs as a **Bluetooth Low Energy (BLE)** device, developers need to ensure that they design their applications to run efficiently so that they don't impact the device's battery. This is very important when designing the custom watch faces or apps that use location service functionality.

The following image describes the architecture between the handheld device and the wearable device. In the next section, we will take a look into some of the wearable APIs that come as part of Google Play services, and explain their purpose when it comes to communicating between the mobile device and the wearable:

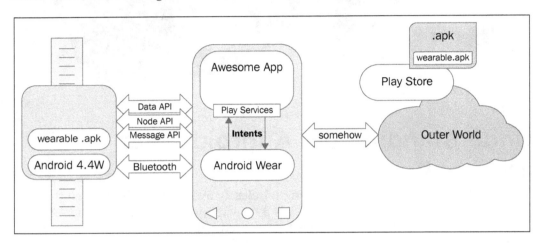

Once the connection is established, you can then start looking at sending and synching data between the two devices. When a connection between two devices has been established, each node can handle any given number of different functions. For example, one node can handle the camera part on the mobile, while another node could keep track of a user's GPS coordinates on the wearable device.

In the following list, we will explain each of the APIs, which are presented in the preceding screenshot. In later chapters, we will be using these in more depth, so at this stage, I will just be providing a brief introduction:

- **Node API**: The `NodeApi` class is responsible for keeping track of all connected or disconnected nodes that have been established within the wearable network by using the `NodeListener` interface method. When a node establishes a connection between the handheld and the wearable, `MessageApi` quietly begins to send a message from the wearable device to the handheld device that it is paired with, which the user is signed in to with their Google account. This sends a notification to the `NodeListener` method that then begins to get information about each node.

- **Message API**: The `MessageApi` class is responsible for sending across short messages to each of the connected network nodes between the wearable and the handheld device. Once a message has been received, a background listener service on the receiving side (`MessageListener`) will be called so that it can get the message.

- **Data API**: The `DataApi` class is responsible for synching data between the connected Android wearable and the handheld device, and takes care of providing the synching mechanism on both sides. In addition to synching data, the big appeal of the data API is that when a user's connection gets disconnected from the paired smartphone, the data will be automatically transferred when the connection is restored, without the user needing to worry about handling data issues. When the data API receives messages from `MessageApi`, a background listening service on the receiving side (`addListener`) will be called as part of the `DataListener` interface method. Once the `addListener` method determines that a change has occurred, a call is made to the `onDataChanged` method.

It is extremely important that you remember to implement `WearableListenerService` on both the Android wearable and the handheld device in order to listen for the events received by `WearableListenerService`.

It is worthwhile to mention that all of the Android Wear APIs are included in Google Play services 5.0. It's important to note that wear itself supports only 4.3 devices and above. This is basically due to the fact that Android Wear requires Bluetooth LE, which is only available in 4.3 and above versions.

In the following chapters of this book, we will be taking a look at how to implement some of these APIs to communicate between our Android device and our wearable device, so stay tuned.

Setting up an Android development environment

In this section, we are going to look at the key concepts to get you started with Android Wear development. Google recommends using Android Studio for development, because of the tight integration it has with the Wear development APIs, as well as the streamlined build cycle with Gradle, that helps minimize a lot of the manual configuration that the developer would need to do in other IDEs.

Before you begin, and as a prerequisite to starting to work with Android Studio, you will need to ensure that your system has the latest version of the **Java Runtime Environment (JRE)** installed for the version of the operating system that you are using.

 To determine if your system has the JRE or the **Java Development Kit (JDK)** installed, open a new terminal window and issue the following command from the command line:

```
java -version
```

Once you have determined if you have Java installed, you can proceed to download Android Studio for your version of the operating system. The Android Studio package can be downloaded from the Android developer tools web page at `http://developer.android.com/sdk/installing/studio.html`.

Android Studio for Windows systems can be downloaded from `https://dl.google.com/dl/android/studio/install/1.2.2.0/android-studio-bundle-141.1980579-windows.exe`.

Android Studio for Mac OS X systems can be downloaded from `https://dl.google.com/dl/android/studio/install/1.2.2.0/android-studio-ide-141.1980579-mac.dmg`.

Android Studio for Linux systems can be downloaded from `https://dl.google.com/dl/android/studio/ide-zips/1.2.2.0/android-studio-ide-141.1980579-linux.zip`.

Now that you have downloaded and installed Android Studio, you can begin installing the Android 4.4W (API 20) for your system:

1. Launch **Android SDK Manager** by using the SDK Manager in Android to download API level 20 (4.4 KitKat Wear).

2. Select and click on the **Android 4.4W.2 (API 20)** package:

You will notice that we have chosen to install the Android Wear system images for both ARM and Intel. Intel delivers greater performance while running your app using the Android Emulator, but you can select the appropriate one for your chipset. If you decide to install both, the Android Studio IDE at design time will inform you which one is supported.

In the next section, we will need to install the Android Wear support libraries for our Android Wear application that will allow your Android wearable app to target a specific version of the Android SDK APIs.

Installing the Android Wear support library

The Android Wear support library contains a set of numerous code libraries that allow you to target a specific version of the Android SDK APIs. Each library contains a different set of features that can help to improve the look of your application, and with the release of Android 5.0 you can incorporate Material design as well as add support for rich notification features.

The benefit of using the latest Android Wear support libraries is that it allows your applications to take advantage of the new and improved features for devices that are running Android 5.0 and above. However, while your app can still run on devices running Android 1.6 and above, some features will not be available:

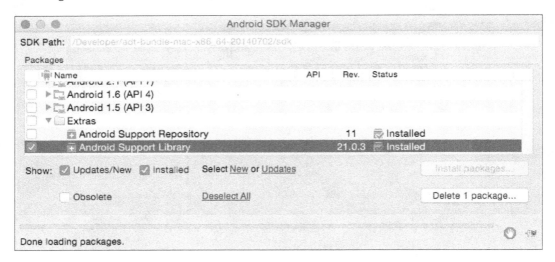

Scroll down to the Extras folder, and select the **Android Support Library**, as shown in the preceding screenshot. Next, click on the **Install packages** button as shown in the preceding screenshot.

If you are using Android Studio for your Android Wear development, this can provide you with a much easier and more convenient way of adding the Android Wear support library to your existing applications. Android Studio uses a module model, where the mobile and wear modules are part of a single project.

Setting up and configuring the Android (AVD) Emulator

In our next step, we will need to set up and configure our Android Wear Emulator. This will allow us to test our Android Wear apps that we will be developing throughout this book.

Open your **Android Virtual Device (AVD)** manager and create a new virtual device for your Android Wear, as shown in the following screenshot:

Before we end this section, it is worth mentioning that while Android Emulator is the most powerful and convenient tool that you will use throughout your development of Android apps, it is important for developers to understand the types of limitations it comes with, which are explained in the following points:

- The Android Emulator simulates real handheld device behavior, but not specific hardware implementations
- Sensor information, such as satellite location, battery, and power settings, as well as network connectivity, is all simulated using your computer
- Access to the camera hardware is not fully functional
- There is no ability to place or receive phone calls, or send SMS messages, as these are all simulated
- There is no support for USB available

As you can see, using the Android emulator is not recommended as a substitute for testing your apps on a true handset or device. Now that we have set up all of the preliminary configurations, we can start to build our Android Wear application.

Building a simple Android wearable application

In this section, we will take a look at how to create a simple Hello World Android Wear application by performing the following steps:

1. Launch **Android Studio**, and then navigate to the **File** | **New Project** option.

2. Next, enter in `HelloAndroidWear` for the **Application name** field.

3. Then provide a name for the **Company Domain** field.

4. Next, choose **Project location** where you would like to save your application code:

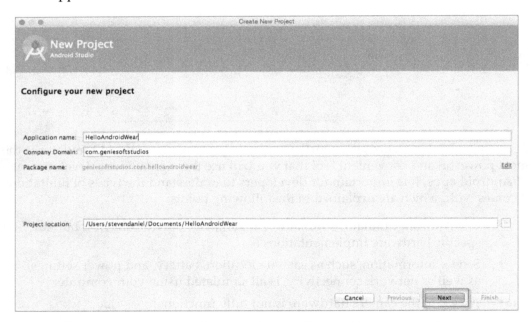

5. Finally, click on the **Next** button to proceed to the next step. On the second wizard screen, we need to specify the form factors using which our application will run. On this screen, we choose the **Minimum SDK** versions for phone and tablet, Android TV, and Android Wear.

6. Click the **Phone and Tablet** option and choose the **API 19: Android 4.4 (KitKat)** option for **Minimum SDK**. Choosing this option allows your application to target more devices that are active on the Google Play Store, with the added cost of having fewer features available for these devices.

7. Next, click on the **Wear** option and choose the **API 20: Android 4.4 (KitKat Wear)** option for **Minimum SDK**:

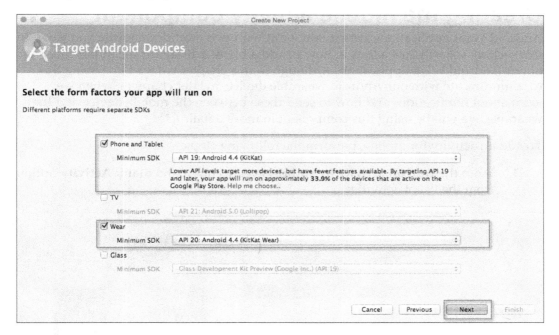

8. Next, click on the **Next** button to proceed to the next step in the wizard.

In our next step, we will be taking a look at how to add a blank activity to our application for its mobile section.

 An activity is basically an application component that provides a screen with which users can interact in order to do something, such as dial the phone, take a photo, send an e-mail, or view a map.

Each activity is given a container to draw its user interface. The container typically fills the screen, but may be smaller than the screen and float on top of other windows.

Creating the mobile activity component

Android Wear applications are actually built with two modules: **mobile** and **wear**. In this section, we will take a look at how to add a blank activity for the mobile portion of our Android Wear application. Although in this chapter we won't be using this to communicate with our Android wearable device, in later chapters, when we learn about notifications and how to send these between the mobile device and the wearable, we will be using this component in more detail.

To add an activity for mobile, perform the following steps:

1. From the **Add an activity to Mobile** screen, choose the **Blank Activity** option from the list of activities:

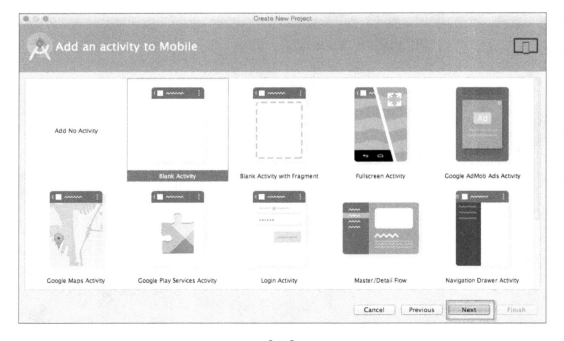

2. Now, click on the **Next** button to proceed to the next step in the wizard.

    In our next step, we need to set up and customize our **Blank Activity** properties that can be used by our application. Here, we specify the name of the activity, layouts, and title, as well as its menu resource name that it will be using (if it contains a menu bar).

3. From the **Customize the Activity** screen, accept the default properties that have been created for you by the wizard:

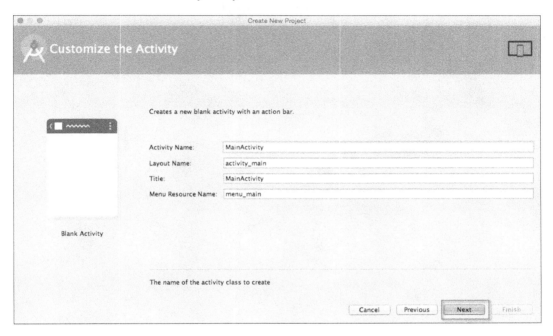

4. Click on the **Next** button to proceed to the next step in the wizard.

Creating the Android Wear activity component

In this section, we will take a look at how to add a blank activity for the wearable portion of our Android Wear application. This will be used to determine how our Android wearable behaves. Once added, this will contain two different watch views: one containing round watch faces and the other containing a square watch look.

To add an activity for Wear, follow these steps:

1. From the **Add an activity to Wear** screen, choose the **Blank Wear Activity** option:

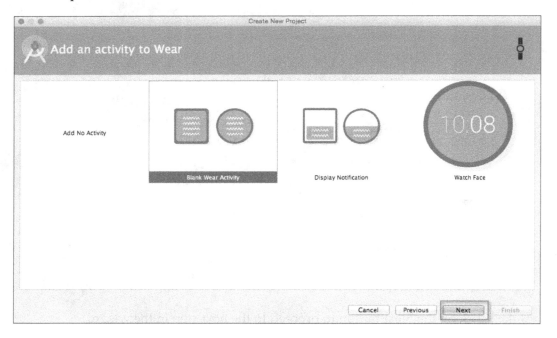

2. Click on the **Next** button to proceed to the final step in the wizard.

    In our final step, we need to customize our Blank Wear Activity properties that can be used by our Android wearable. Here, we specify the name of the activity as well as the layouts for the watch faces for round and rectangle.

3. From the **Customize the Activity** screen, accept the default properties that have been created for you by the wizard:

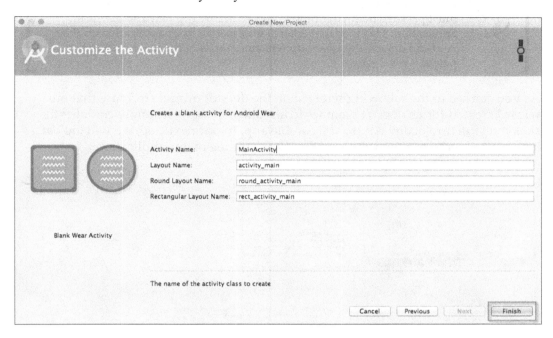

4. Next, click on the **Finish** button to proceed, and your Android wearable project will be generated for you. After a few moments, the Android Studio window will be displayed with your project loaded.

When you take a look at the project that the wizard created for you, the first thing you will notice is that it contains two modules: mobile and wear. The mobile portion of our project is the application that will run on a phone device that will be used to communicate with our wearable device. The wear portion is the application that will be installed on the Android wearable device.

When developing Android wearable applications, these cannot simply be packaged separately and uploaded to the Google Play Store. You must package both your mobile and wearable app into a single APK application. This is so that when a user installs your application on their Android device, the wear app will automatically be transferred to the paired wear device.

As you can see in the following screenshot, the default project structure that our wizard created for us doesn't seem to do anything special—it simply contains the skeleton structure of any Android wearable app. In our next step, we will look at how we can write our own code for the **Wear** module of our application:

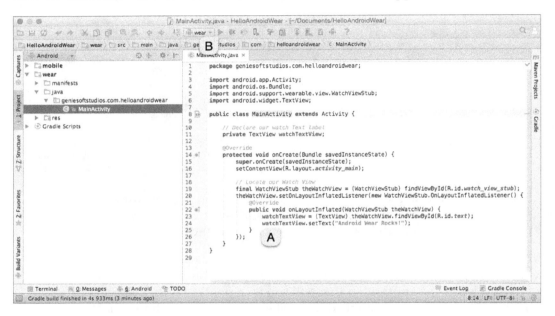

In our next step, we will begin modifying our `MainActivity` class file to display our custom Android wearable welcome message to the user within the wearable watch face :

1. Open the **HelloAndroidWear** project from within our **Project Navigator** window.

2. From the **Project Navigator** window, expand the **Wear** section, select, and expand the **java** section.

3. Modify the following code in the MainActivity.java file of the project:

```
private TextView watchTextView;
@Override
protected void onCreate(Bundle savedInstanceState) {
   super.onCreate(savedInstanceState);
   setContentView(R.layout.activity_main);
   final WatchViewStub theWatchView = (WatchViewStub)
   theWatchView.setOnLayoutInflatedListener(new
     WatchViewStub.OnLayoutInflatedListener() {
     @Override
     public void onLayoutInflated(WatchViewStub
       theWatchView) {
       watchTextView = (TextView)
         theWatchView.findViewById(R.id.text);
       watchTextView.setText("Android Wear Rocks!");
     }
   });
}
```

Downloading the example code

You can download the example code files for all Packt books you have purchased from your account at http://www.packtpub.com. If you purchased this book elsewhere, you can visit http://www.packtpub.com/support and register to have the files e-mailed directly to you.

In the preceding code snippet, we start by creating a new activity, which is handled by the onCreate method. This method is responsible for starting the activity when the application is launched and the emulator sets up the content, prior to displaying the watch layouts on the screen using the setContentView method.

Next, we declare an instance of WatchViewStub, which is used to detect the specific watch type being used at runtime and allows you to inflate a rectangular or round layout. Since we cannot access these child views until inflation has completed, we implement the OnLayoutInflatedListener interface to handle this, which allows us to get a reference to the child views by using the findViewById class. Once we have established this reference, we can then proceed to update the watchTextView with our **Android Wear Rocks!** text.

When we make a call to invoke the onLayoutInflated method from inside our WatchViewStub class, this will begin to load the corresponding layout resource for either the rect_activity_main. xml file for our square view or the round_activity_main.xml file for our round watch views. Once the view has inflated within its parent view, this will then get added to the view hierarchy chain of your application prior to making it visible.

If you would like to learn more about the activity lifecycle and the different states that it takes on, you can refer to the documentation at http://developer.android.com/reference/android/app/ Activity.html#ActivityLifecycle.

Now, we can finally compile, build, and run our application. Click on the green button labeled **B** in the preceding screenshot or simply press *CMD + F9*, and choose **Android Virtual Device** from the list of Android Emulators:

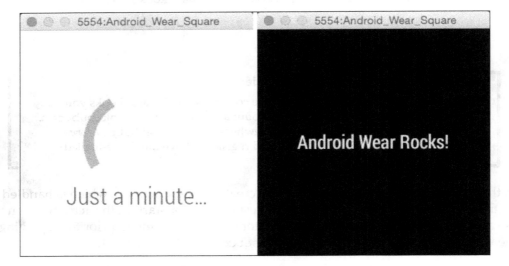

Once the emulator has launched and after a few moments, you should see your app displayed, as shown in the preceding screenshot.

Summary

In this chapter, we explored the features and key concepts of the Android Wear platform; we discussed about the Android Wear architecture; and then we moved on to look at how to configure our Android environment, setting up the Android SDK and AVD, before finally taking a look at how to build our simple Android wearable app.

In the next chapter, we will cover the features of Android notifications and explore the different ways that we can send notifications between the handheld device and the Android wearable to create basic and custom notifications. We will learn how we can use the Android voice capabilities to allow the user to respond to notification messages by using just their voice, and how we can use a method called page stacking to receive multiple notifications.

2

Creating Notifications

This chapter provides you with a background of essential features of Android notifications and how we can use these to send messages between a handheld device and a wearable device.

We will be taking a look at how we can use different notification methods to develop an application that shows how we can create basic and custom notifications. You will also learn how you can incorporate and make use of Android's voice capabilities to respond to notification messages that are contained within a notification.

At the end of the chapter, we will take a look at how we can use notification stacking to display multiple notification messages within a notification.

This chapter includes the following topics:

- Introducing Android notifications
- Creating a basic notification for wearables
- Creating a custom notification for wearables
- Receiving voice input within a notification
- Receiving multiple notifications through a process called page stacking

Introducing Android notifications

Android notification is basically a way of communicating with the user. This is done by letting the user know of an upcoming appointment or of an incoming call or SMS message. The user can then decide how to respond to the option presented to him/her.

Notifications in wear are the result of events that happen on the paired smartphone, which are then mirrored on the wearable device, or contextual events like location-aware events or time, and date-based reminders. Android notification messaging works by communicating between the handheld device and wearable device over Bluetooth. When the connection has been established, the Google Play Service notification messages can be exchanged between the handheld device and the wearable. An example of a notification message that has been sent from the Android handheld device to the Android wearable can be seen in the following screenshot:

In the next sections, we will be taking a look at the different ways we can send notifications and how we can respond to them.

Creating a basic notification for wearables

In this section, we will take a look at how to create a basic notification that will be displayed on our wearable device. So let's get started!

Firstly, create a new project in Android Studio by following these simple steps:

1. Launch **Android Studio**, and then navigate to the **File | New Project** menu option.

2. Next, enter `WearNotifications` for the **Application name** field.

3. Then, provide the name for the **Company Domain** field.

4. Next, choose **Project location** where you would like to save your application code:

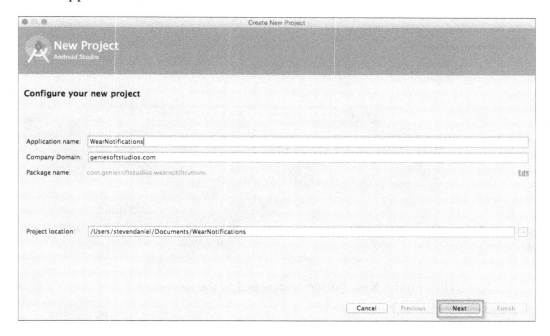

5. Finally, click on the **Next** button to proceed to the next step.

Specifying the form factors

Next, we will need to specify the form factors that our application will use, as will be using a single module instead of the app and wear modules as discussed previously in *Chapter 1, Understanding Android Wearables and Building Your First Android Wear App.*

On the following screen we will need to choose the minimum SDK version for our phone/tablet and perform the following steps:

1. Click on the **Phone and Tablet** option and choose the **API 19: Android 4.4 (KitKat)** option for **Minimum SDK**:

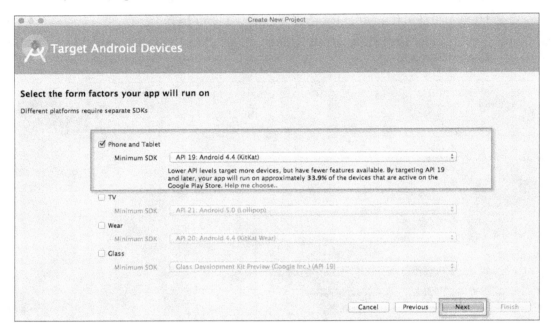

2. Next, click on the **Next** button to proceed to the next step.

Adding and customizing a blank activity

In our next step, we will need to add **Blank Activity** to our application project for the mobile section of our app. If you remember from *Chapter 1, Understanding Android Wearables and Building Your First Android Wear App*, we had mentioned that an activity is basically an application component that provides a screen that the users can interact with. Perform the following steps to add and customize a blank activity:

1. From the **Add an activity to Mobile** screen, choose the **Blank Activity** option from the list of activities shown and click on the **Next** button to proceed to the next step:

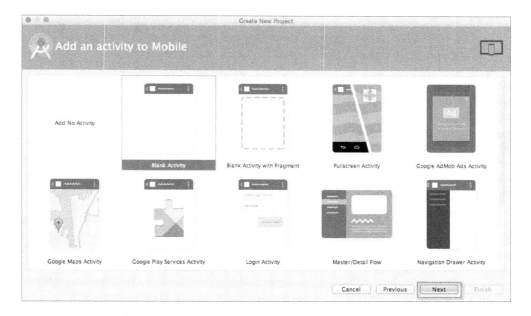

Next, we need to customize the properties for **Blank Activity** so that it can be used by our application. Here we will need to specify the name of our activity, the layout information, and the title as well as its menu resource file.

2. From the **Customize the Activity** screen, accept the default properties that the wizard has created for us:

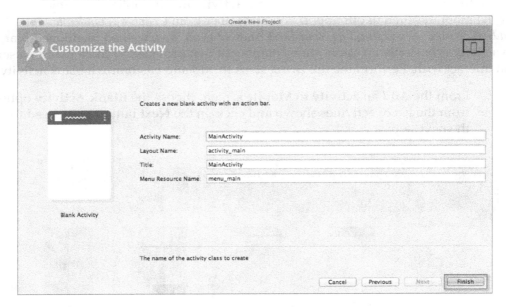

3. Click on the **Finish** button. Your project will be generated by the wizard. After a few moments, the **Android Studio** window will appear with your project displayed.

Adding dependencies to Gradle scripts

In our next step, we need to add a new dependency to the **Gradle Scripts** section of our project.

This will provide us with the ability to send notification messages between the handheld and the wearable device:

1. From the **Project Navigator** window, double-click on the **build.gradle (Module: app)** node, which is located within the **Gradle Scripts** section of **Project Navigator,** as shown in the following screenshot:

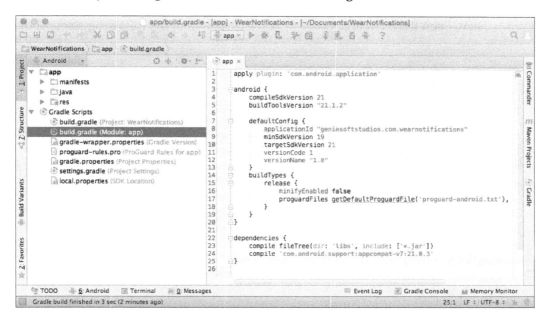

Gradle is the packaging tool that comes bundled with Android Studio, and it takes care of building the application's APK package. Gradle supports incremental builds and intelligently determines what parts of the project source code are up to date, without the need to recompile the whole project.

2. Next, under the **dependencies** section, add the following highlighted code:

```
dependencies {
    compile fileTree(dir: 'libs', include: ['*.jar'])
    compile 'com.android.support:appcompat-v7:21.0.3'
    compile 'com.android.support:support-v4:20.0.+'
}
```

3. Then, open the `MainActivity.java` file as shown in the following screenshot:

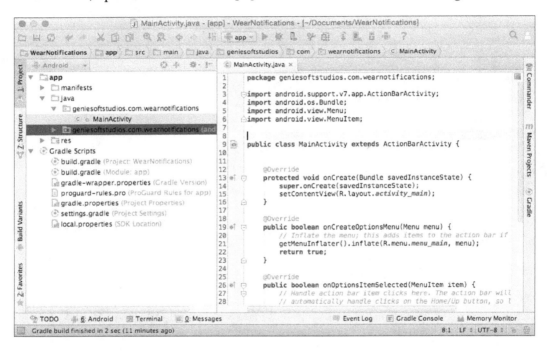

4. Next, modify the import statements within our `MainActivity.java` file with the highlighted entries shown in the following code snippet:

```
import android.app.Notification;
import android.app.PendingIntent;
import android.support.v4.app.NotificationCompat;
import android.support.v4.app.NotificationManagerCompat;
import android.content.Intent;
import android.support.v7.app.ActionBarActivity;
```

```
import android.os.Bundle;
import android.view.Menu;
import android.view.MenuItem;
// Handling Custom Notifications
import android.view.View;
import android.widget.Button;
import android.widget.EditText;


// Handling Voice Notifications
import android.support.v4.app.RemoteInput;
import android.util.Log;
```

5. Next, from the **Project Navigator** window, open the `strings.xml` file, which is located in the `res | values` folder, as follows:

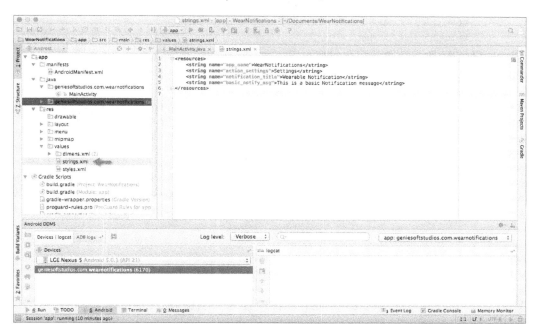

6. Next, add the following highlighted entries as shown in the following code snippet:

```
<resources>
<string name="app_name">WearNotifications</string>
<string name="action_settings">Settings</string>
<string name="notification_title">Wearable
      Notification</string>
<string name="basic_notify_msg">This is a basic
      Notification message</string>
<string name="notification_button">Send
      Notification</string>
<string name="notification_message">Custom notification
      message</string>
<string-array name="voice_choices">
<item>OK</item>
<item>Great</item>
<item>Im not sure</item>
</string-array>
</resources>
```

7. Next, we need to add our notification message ID to the `MainActivity` class. This value can be any arbitrary string value, but each unique notification will need to have it's own ID as it will be responsible for sending notification message between our handheld device and the wearable. This is shown in the following code snippet:

```
public class MainActivity extends ActionBarActivity {

    // Set up our Notification message ID
    int NOTIFICATION_ID = 001;
```

8. Then, modify the `onCreate` method as follows:

```
@Override
protected void onCreate(Bundle savedInstanceState)
{
    super.onCreate(savedInstanceState);
    setContentView(R.layout.activity_main);

    // Clear all previous notifications and
    // generate new unique ids
```

```
NotificationManagerCompat.from(this).cancelAll();

        Intent intent = new Intent(this, MainActivity.class);
        PendingIntent pendingIntent = PendingIntent.
getActivity(this, 0,
intent,
PendingIntent.FLAG_UPDATE_CURRENT);

        // Method to display our basic notification
        displayBasicNotification(pendingIntent);
    }
```

9. Next, we need to create a new `displayBasicNotification` method that will be responsible for sending a basic notification message to the Android wearable device, which is shown in the following code snippet:

```
// Method for displaying our basic notification message
public void displayBasicNotification(PendingIntent pendingIntent)
    {
// Set up our Notification Action class method
        NotificationCompat.Action action = new
NotificationCompat.Action.Builder(
R.mipmap.ic_launcher,
getString(R.string.notification_title),
pendingIntent)
.build();

Notification notification = new
NotificationCompat.Builder(MainActivity.this)
.setContentText(getString(R.string.basic_notify_msg))
.setContentTitle(getText(R.string.notification_title))
                .setSmallIcon(R.mipmap.ic_launcher)
                .extend(new
NotificationCompat.WearableExtender().addAction(action))
                .build();

NotificationManagerCompat notificationManagerCompat =
NotificationManagerCompat.from(MainActivity.this);
notificationManagerCompat.notify(NOTIFICATION_ID, notification);
    }
```

In the preceding code snippet, we started by adding our import statements that are responsible for handling the notifications between our Android handheld device and the wearable device. Then we added a NOTIFICATION_ID constant variable that is a unique identifier for our notification and will be responsible for sending the notification between the handheld and the wearable device. Next, we call the cancelAll() method to our NotificationManagerCompat class that cancels all of the previously shown notifications prior to creation of the two class variables, Intent and PendingIntent.

Intent is used to bind itself to the current activities connection at runtime to send messages from one component to another within the same application context. PendingIntent is basically a token that we provide to NotificationManager that provides our application permission to perform operation, for the application's current activity.

Then, specify FLAG_UPDATE_CURRENT on our PendingIntent class object to indicate that pendingIntent already exists. This is basically to keep our pendingIntent active so that we can replace the intent with the code declared by the new Intent class and then the pendingIntent object is passed as a parameter to our displayBasicNotification method.

Within our displayBasicNotification method, we begin by declaring a NotificationCompat.Action action object that basically builds our notification that will be shown as part of the notification and must include an icon, label, and PendingItent that will be fired when the user selects the action.

We create a Notification object and a NotificationCompat.Builder class for the main activity, and pass our notification message to the setContentText class and specify its title and icon. We use the extend property to add support for wearable devices and pass our action variable so that actions can be responded to by these devices. Next, we use .build to construct our notification object and then specify NotificationManagerCompat that accepts a .from property. This contains our MainActivity class instance context, a .notify property that contains the ID of our notification as specified by NOTIFICATION_ID, as well as the notification object that will be used to post the notification to the system.

Next, we can finally compile, build, and run our application. Simply press *CMD + F9* and choose your AVD or your Android handheld device from the list of Android Emulators.

Once the emulator has launched, you should see the notification displayed in the following screenshot:

As you can see, creating notifications is quite simple. In the next section, we will take a look at how we can send a custom notification message entered by the user.

Creating a custom notification for wearables

In this section, we will take a look at how we can create a custom notification message that will be entered by the user within our code example. This will be done on the paired smartwatch, with the notification being displayed on the Android wearable device.

So let's get started:

1. First, open the `activity_main.xml` file, which is located in the `res | layout` folder structure within **Project Navigator,** and add the highlighted entries as shown in the following code snippet:

```
<RelativeLayout
    xmlns:android="http://schemas.android.com/apk/res/android"
    xmlns:tools="http://schemas.android.com/tools"
    android:layout_width="match_parent"
    android:layout_height="match_parent"
    android:paddingLeft="@dimen/activity_horizontal_margin"
    android:paddingRight="@dimen/activity_horizontal_margin"
    android:paddingTop="@dimen/activity_vertical_margin"
    android:paddingBottom="@dimen/activity_vertical_margin"
    tools:context=".MainActivity">

<Button
        android:layout_width="wrap_content"
        android:layout_height="wrap_content"
        android:text="Send notification"
        android:id="@+id/sendNotificationButton"
        android:layout_centerVertical="true"
android:layout_centerHorizontal="true" />

<EditText
        android:layout_width="wrap_content"
        android:layout_height="wrap_content"
        android:id="@+id/customNotificationInput"
        android:minEms="10"
        android:gravity="center_horizontal"
        android:singleLine="true"
        android:layout_above="@+id/sendNotificationButton"
android:layout_centerHorizontal="true" />

</RelativeLayout>
```

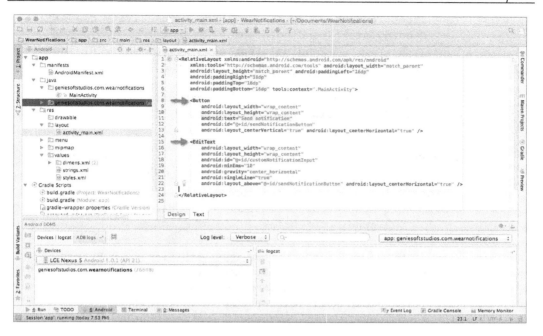

Now that we have created UIs for our application, we can proceed and start to create the additional code that will hook up our button and text field controls to their associated method events.

2. Open the `MainActivity.java` file from within the **Project Navigator** window.

3. Modify the `onCreate` method with the following highlighted code:

```
@Override
    protected void onCreate(Bundle savedInstanceState) {
        super.onCreate(savedInstanceState);
        setContentView(R.layout.activity_main);

        // Clear all previous notifications and
        // generate new unique ids
        NotificationManagerCompat.from(this).cancelAll();
```

```
Intent intent = new Intent(this, MainActivity.class);
PendingIntent pendingIntent =
            PendingIntent.getActivity(this, 0, intent,
            PendingIntent.FLAG_UPDATE_CURRENT);

    // Method to handle our custom notifications
displayCustomNotification(pendingIntent);
    }
```

4. Next, create the `displayCustomNotification` method underneath the `displayBasicNotification` method and add the following code:

```
    // Method for displaying our custom notification message
    public void displayCustomNotification(final PendingIntent
            pendingIntent) {

        // Get a reference to our Send Notifications Button
final Button mSendNotificationButton = (Button)
        findViewById(R.id.sendNotificationButton);
final EditText mSendNotificationInput = (EditText)
        findViewById(R.id.customNotificationInput);

        // Set our notification input hint message and
        // update the text for our button
    mSendNotificationInput.setHint(R.string.notification_message);
    mSendNotificationButton.setText(R.string.notification_button);

        // Set up our Send Notifications Button OnClick method
        mSendNotificationButton.setOnClickListener(new
            View.OnClickListener() {
            @Override
            public void onClick(View v) {
// Get a pointer to our entered in message textBox
                String message =
                mSendNotificationInput.getText().toString();

                // Set up our Notification Action class method
                NotificationCompat.Action action = new
                    NotificationCompat.Action.Builder(
                    R.mipmap.ic_launcher,
```

```
                                getString(R.string.notification_title),
                                pendingIntent)
                                .build();

                        Notification notification = new
                        NotificationCompat.Builder(MainActivity.this)
                        .setContentText(message)
                        .setContentTitle(getText(R.string.notification_title))
                        .setSmallIcon(R.mipmap.ic_launcher)
                        .extend(new
                     NotificationCompat.WearableExtender().addAction(action))
                        .build();
                        NotificationManagerCompat notificationManagerCompat =
                        NotificationManagerCompat.from(MainActivity.this);
                        notificationManagerCompat.notify(NOTIFICATION_ID,
                        notification);
                        }
                });
        }
```

In the preceding code snippet, we started by calling the `cancelAll()` method to our `NotificationManagerCompat` class to clear all of the previous notifications prior to the creation of our class variables, `Intent` and `PendingIntent`. Next, we passed the current class's `PendingIntent` object to our `displayCustomNotification` method that will be responsible for displaying the custom message entered by the user.

Within our `displayCustomNotification` method, we begin by declaring two objects, `mSendNotificationButton` and `mSendNotificationInput`. These objects contain the reference to `Button` and `EditTextField` that will hold the text entered by the user, and we proceed to set some properties for each of these fields.

Next, we set up a `setOnClickListener` object that will handle response to the events when the user has tapped on the **SEND NOTIFICATION** button. Then we specify our `NotificationCompat.Action action` object that basically builds our notification object that will be shown as part of the `notification` service and must include an icon, label, and `PendingItent` that will be fired when the user selects the action.

We create a `Notification` object and a `NotificationCompat.Builder` class for the main activity, pass the message entered by the user to the `setContentText` class, and specify its title and icon. We use the `extend` property to add support for wearable devices, and pass our action variable so that actions can be responded to by these devices.

Next, we use `.build` to construct our notification object and then specify `NotificationManagerCompat` that accepts a `.from` property that contains our `MainActivity` class instance context, a `.notify` property that contains the ID of our notification specified by `NOTIFICATION_ID`, as well as the notification object that will be used to post the notification to the system.

Next, we can finally compile, build, and run our application. Simply press *CMD + F9* and choose your AVD or Android handheld device from the list of Android Emulators:

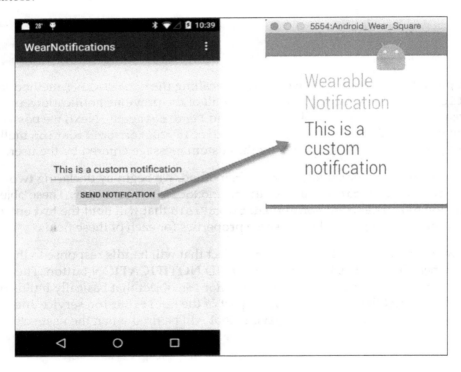

Once the emulator has launched, you should see the custom notification message displayed.

Enter some text within the text field using your Android handheld device, and then click on the **SEND NOTIFICATION** button. After a couple of seconds, you should see the custom notification you entered on your Android wearable and the handheld device.

Receiving voice input within a notification

In this section, we will be taking a look at how we can use notifications that can allow the user to respond to these by simply using their voice. Since Android wearable devices don't contain a keyboard, a user can swipe on a notification and respond to the action using their voice, or simply by choosing from a list of options presented to them, and tapping on the item right from their Android wearable device.

To receive voice input within a notification, perform the following steps:

1. Open the `MainActivity.java` file from within the **Project Navigator** window.

2. Modify the `onCreate(Bundle savedInstanceState)` method and add the following highlighted code:

```
@Override
    protected void onCreate(Bundle savedInstanceState) {
        super.onCreate(savedInstanceState);
        setContentView(R.layout.activity_main);

        // Clear all previous notifications and
        // generate new unique ids
        NotificationManagerCompat.from(this).cancelAll();

        Intent intent = new Intent(this, MainActivity.class);
        PendingIntent pendingIntent =
                PendingIntent.getActivity(this, 0, intent,
                PendingIntent.FLAG_UPDATE_CURRENT);

    // Method to handle voice notifications
    voiceNotifications(pendingIntent);
    }
```

3. Next, create the `voiceNotifications` method underneath the `displayCustomNotification` method and add the following code:

```java
// Method for handling our Voice Notifications
    public void voiceNotifications(PendingIntent pendingIntent)
    {
// Key for the string that's delivered in
        // the action's intent
        final String EXTRA_VOICE_REPLY = "extra_voice_reply";
        final String voiceOptions = "Choose one of these
options";
        String[] voiceChoices =
getResources().getStringArray(R.array.voice_choices);

        final RemoteInput remoteInput = new
RemoteInput.Builder(VOICE_NOTIFY_OPTIONS)
                .setLabel(voiceOptions)
                .setChoices(voiceChoices)
                .build();
// Call our voice notification
        handleVoiceNotifications(remoteInput, pendingIntent);

        // Get the users spoken voice message and display it
        CharSequence replyText = getMessageText(getIntent(),
                EXTRA_VOICE_REPLY);
        if(replyText != null) {
            Log.d("VoiceNotifications", "You replied: " +
            replyText);
        }
    }
```

4. Next, create the `handleVoiceNotifications` method underneath the `voiceNotifications` method and add code as follows:

```java
// Method for responding to Voice Notification messages
    public void handleVoiceNotifications(RemoteInput
remoteInput, PendingIntent pendingIntent)
    {
// Get a reference to our entered in message textBox
String message = "Please respond to this message";
```

```
// Set up our Notification Action class method
NotificationCompat.Action action = new NotificationCompat.Action.
Builder(
                R.mipmap.ic_launcher,
getString(R.string.notification_title),
pendingIntent)
                .addRemoteInput(remoteInput) // Voice Input
                .build();

Notification notification = new
NotificationCompat.Builder(MainActivity.this)
.setContentText(message)
.setContentTitle(getText(R.string.notification_title))
.setSmallIcon(R.mipmap.ic_launcher)
.extend(new
NotificationCompat.WearableExtender().addAction(action))
.build();
NotificationManagerCompat notificationManagerCompat =
NotificationManagerCompat.from(MainActivity.this);
        notificationManagerCompat.notify(NOTIFICATION_ID,
notification);
    }
```

5. Next, create the `getMessageText` method underneath the `handleVoiceNotifications` method and add the following:

```
// Method that accepts an intent and returns the voice
// response, which is referenced by the EXTRA_VOICE_REPLY
// key.
private CharSequence getMessageText(Intent intent,
                        String EXTRA_VOICE_REPLY) {
    Bundle remoteInput =
    RemoteInput.getResultsFromIntent(intent);
    if (remoteInput != null) {
        return
        remoteInput.getCharSequence(EXTRA_VOICE_REPLY);
    }
    return null;
}
```

In the preceding code snippet, we started by calling the `cancelAll()` method to our `NotificationManagerCompat` class to clear all of the previous notifications prior to the creation of class variables `Intent` and `PendingIntent`. Next, we added our method call to `voiceNotifications` that will be responsible for setting up our voice notifications. In this method, we begin by creating our `EXTRA_VOICE_REPLY` string variable and our `voiceChoices` string array that will act as our group container and contain our list of valid responses that the user can respond to using their voice. In the next step, we begin by creating our `RemoteInput.Builder` class method that will hold our label and list of choices that the user can respond to, by either using their voice or tapping on them with their finger. This method will then display the spoken text on the console window.

Next, we declare our `handleVoiceNotifications` method that will be responsible for handling the voice notifications. This method accepts our `remoteInput` and `pendingIntent` class as parameters and begins to declare a `NotificationCompat.Action action` object. This basically builds our notification object that will be shown as part of the `notification` service and must include an icon, label, and `PendingItent` that will be fired when the user selects the action. One different thing you will notice here is that we added a new property called `addRemoteInput` and passed our `remoteInput` instance variable to this. This is done to collect the input from the user when the response has been sent. Next, we start by creating our `notificationManager` object based on our `NotificationManagerCompat` instance class that accepts the `.from` property. This contains our current class instance context, which is referred to as `this`. It also contains a `.notify` property that contains the ID of our voice notification as specified by `NOTIFICATION_ID`, as well as our notification object that will be used to post the notification to the system.

Finally, we create our `getMessageText` method that accepts the intent of our `Activity` class that is carrying the information from the wearable, and returns back the voice response. This is referenced by our `EXTRA_VOICE_REPLY` key back to our `voiceNotifications` method to display the captured voice response to the user.

Next, we can finally compile, build, and run our application. Simply press *CMD + F9* and choose your AVD or Android handheld device from the list of Android emulators:

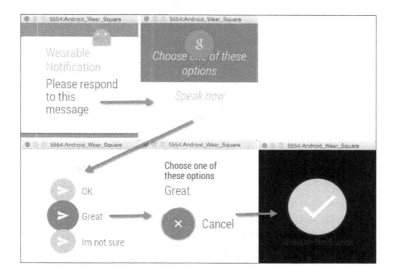

Once the emulator has launched, you should see the notification messages displayed and grouped as shown in the preceding screenshot.

Receiving multiple notifications through a process called page stacking

In the previous sections, we have looked at how we can use notifications to send basic and customized notifications to our Android wearable device. In the final section, we will look at how we can use our notification service to group notifications through a process called page stacking.

This process helps us to intelligently collate and bundle similar items within a single collection to provide us with easy navigation from the same app. So, let's get started:

1. Open the `MainActivity.java` file from within our **Project Navigator** window.

2. Modify the `onCreate(Bundle savedInstanceState)` method and add the following highlighted code:

```
@Override
    protected void onCreate(Bundle savedInstanceState) {
        super.onCreate(savedInstanceState);
        setContentView(R.layout.activity_main);

        // Clear all previous notifications and
        // generate new unique ids
        NotificationManagerCompat.from(this).cancelAll();

        Intent intent = new Intent(this, MainActivity.class);
        PendingIntent pendingIntent =
PendingIntent.getActivity(this, 0, intent,
PendingIntent.FLAG_UPDATE_CURRENT);

    // Method to display our Page-Stacking notifications
    displayPageStackNotifications();
        }
```

3. Next, create the `displayPageStackNotifications` method underneath the `handleVoiceNotifications` method and add the following code:

```
// Method for displaying our page stacking
    // notification messages
    public void displayPageStackNotifications() {

        int stackNotificationId = 0;
        int MAX_NOTIFICATIONS = 2;

        // String to represent and group all of the notifications
        // that will be a part of it.
```

```java
final String GROUP_NOTIFICATIONS = "group_notifications";

// Group notification that will be visible on the phone
Notification summaryNotification = new
        NotificationCompat.Builder(this)
        .setContentTitle(MAX_NOTIFICATIONS + "
        Notifications Received")
        .setContentText("You have received " +
        MAX_NOTIFICATIONS + " messages")
        .setSmallIcon(R.mipmap.ic_launcher)
        .setGroup(GROUP_NOTIFICATIONS)
        .setGroupSummary(true)
        .build();

// Create our first view Intent
Intent viewIntent1 = new Intent(this, MainActivity.class);
PendingIntent viewPendingIntent1 =
        PendingIntent.getActivity(this,
        (stackNotificationId + 1), viewIntent1, 0);

Notification notification1 = new
        NotificationCompat.Builder(this)
        .addAction(R.mipmap.ic_launcher, "Sounds Great",
        viewPendingIntent1)
        .setContentTitle("Movie Message")
        .setContentText("Do you want to go to the
        movies?")
        .setSmallIcon(R.mipmap.ic_launcher)
        .setGroup(GROUP_NOTIFICATIONS)
        .build();

// Create our second view intent
Intent viewIntent2 = new Intent(this, MainActivity.class);
PendingIntent viewPendingIntent2 =
        PendingIntent.getActivity(this,
        (stackNotificationId + 2), viewIntent2, 0);
```

```
Notification notification2 = new
        NotificationCompat.Builder(this)
        .addAction(R.mipmap.ic_launcher, "Why not",
        viewPendingIntent2)
        .setContentTitle("Red Wine Message")
        .setContentText("Another glass of Red Wine?")
        .setSmallIcon(R.mipmap.ic_launcher)
        .setGroup(GROUP_NOTIFICATIONS)
        .build();

// Issue our group notification message
NotificationManagerCompat notificationManager =
        NotificationManagerCompat.from(this);
notificationManager.notify(stackNotificationId,
        summaryNotification);
// Then, issue each of our separate wearable notifications
notificationManager.notify((stackNotificationId + 1),
                            notification1);
notificationManager.notify((stackNotificationId + 2),
                            notification2);

}
```

In the preceding code snippet, we started by adding our import statements that are responsible for handling the notifications between our Android handheld device and the wearable device, just like how we created the basic and custom notifications.

Just like our previous examples in this chapter, we make a call to the cancelAll() method in our NotificationManagerCompat class to clear all of the previous notifications prior to the creation of our class variables, Intent and PendingIntent. Next, we add the method call to displayPageStackNotifications that will be responsible for displaying our page stacking notifications. In this method, we begin by initializing stackNotificationId, which will be used to keep a track of each notification that we create. We then create a MAX_NOTIFICATIONS variable that contains the total number of page stacking notifications, and then create a GROUP_NOTIFICATIONS string that contains our group container to hold each of our notifications.

In our next step, we begin by creating our group notification class using our NotificationCompat.Builder class method as we did in the previous sections, but with one difference, that is we include the setGroup and setGroupSummary properties.

The setGroup property groups all of your notifications in a single card, and then allows you to scroll through each of these on your wearable device.

The setGroupSummary property allows you to provide a summary as part of your page stacking notification, for example, it shows how many unread messages you have. In the next step, we start by creating our view that will be used for our first notification and get the intent and pendingIntent current classes. Then we increment stackNotification for our first view and begin setting up our notification adding it as part of our GROUP_NOTIFICATIONS group. After this, we do the same for our next notifications view and increment the number by one each time, so that we do not override the previous notifications intent view before adding this item as part of our group.

Finally, we create our notificationManager object based on the instance of the NotificationManagerCompat class that accepts the .from property. This contains our current class instance context, which is referred to as this. It also contains a .notify property that has the ID of our stack notification as specified by stackNotificationId, as well as our summaryNotification object that will be used to post the notification to the system. Then we begin issuing each of our separate wearable notifications using our notificationManager.notify method and passing stackNotificationId to our notification service, as well as the notification that we want to start posting.

Next, we can finally compile, build, and run our application. Simply press *CMD + F9* and choose your AVD or your Android handheld device from the list of android emulators.

Once the emulator has launched, you should see the notification messages displayed and grouped as shown in the following screenshot:

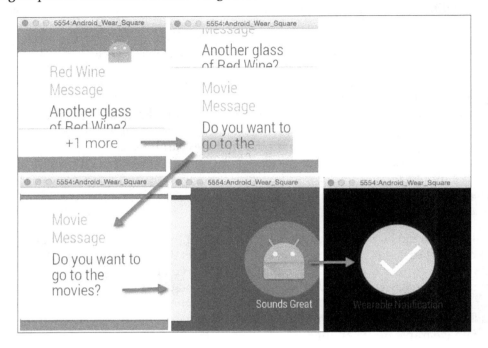

The preceding screenshot shows the current notification workflow including each of the respective screens when the user responds to the actions.

 Android Wear often displays a confirmation screen after user input, which can be programmatically manipulated, either with an animation or a timer-based confirmation that lets the user cancel their selection. To obtain further information about this process, it is worth checking out the *Showing Confirmations* documentation at `https://developer.android.com/training/wearables/ui/confirm.html`.

Summary

In this chapter, we explored the various ways by which we are able to send notifications between the handheld device and the Android wearable. We looked at how to build a simple Android handheld app with customized notifications for Android Wear to show different types of notifications. Next, we looked at how to send basic and custom notification messages, which can be customized by the user. Also, we looked at how we can use notifications that can be responded to by the user using their voice.

In the final part of this chapter, we looked at how we can receive multiple notifications on our Android wearable, and group these by using a process called page stacking.

In the next chapter, we will gain an understanding of how we can build effective UIs to create, customize, and draw watch faces using Google's official API that will provide us the ability to present information within the watch area.

3
Creating, Debugging, and Packaging Wearable Apps

This chapter will provide you with the background and understanding of how you can build and package your own apps for Android Wear. We will learn how we can draw our own Android watch face using Google's official API, and also to present information within the watch area.

We will learn how to create a custom class that inherits from Google's latest watch face service API that will enable us to respond to watch face events to handle screen updates, draw watch face elements, and respond to changes between the interactive and ambient modes.

Finally, we will learn how we can effectively debug an Android wearable application over Bluetooth, and then take a look at the Android design principles to ensure that our application conforms to these. Then we will move on to learn how to package an Android wearable application so that it can communicate and be used by our handheld device.

This chapter includes the following topics:

- Creating a custom Android Wear watch face service class
- Presenting information inside a custom watch face
- Using Bluetooth to debug your Android wearable app
- Running your app directly on the Android wearable device
- Introducing the Android wearable user interface guidelines
- Packaging your Android wearable app within a handheld device

Creating an Android wearable watch face app

Prior to Google releasing the official Android watch face API in December 2014, developers had to find alternative ways to present information inside the watch face layout. In this section, we will take a look at the steps required to create a custom watch face service that will enable us to communicate with our watch and present information within the watch face area on the home screen.

Firstly, create a new project in Android Studio by following these simple steps:

1. Launch Android Studio, and then navigate to the **File | New Project** menu option.

2. Next, enter in `CustomWatchFace` for the **Application name** field.

3. Then, provide the name for the **Company Domain** field.

4. Next, choose **Project location** where you would like to save your application code:

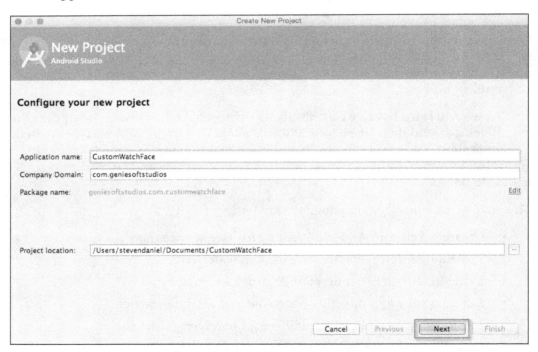

5. Click on the **Next** button to proceed to the next step.

   Next, we will need to specify the form factors for our phone/tablet and Android Wear devices that our application will run on. On this screen, we will need to choose the minimum SDK version for our phone/tablet and Android Wear.

6. Click on the **Phone and Tablet** option and choose **API 19: Android 4.4 (KitKat)** for **Minimum SDK**.

7. Click on the **Wear** option and choose the **API 21: Android 5.0 (Lollipop)** option for **Minimum SDK** as we want to use the watch face service API:

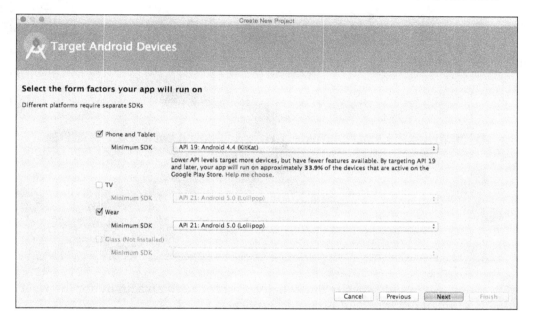

8. Click on the **Next** button to proceed to the next step.

   In the next step, we need to specify that we don't want to add an activity for both the mobile and wear sections of our application.

9. From the **Add an activity to Mobile** screen, choose the **Add No Activity** option from the list of activities shown, and click on the **Next** button to proceed to the next step:

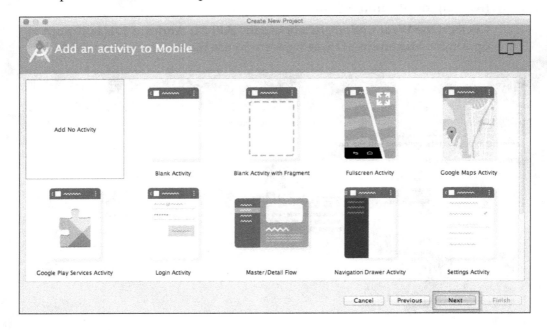

Next, we need to specify that we don't want to add an activity for the wear section of our Android wearable application.

10. From the **Add an activity to Wear** screen, choose the **Add No Activity** option from the list of activities shown, and click on the **Finish** button:

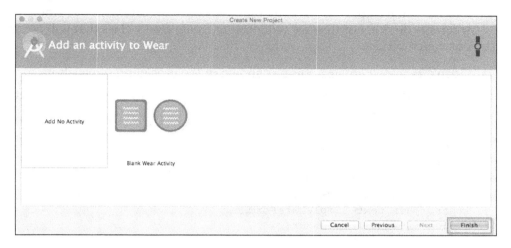

At this point, the wizard will generate your project, and after a few moments the Android Studio window will appear with your project displayed in it. In the next section, we proceed by creating a custom Android `WatchFace` class that will be used to present content and update the watch face user interface.

Presenting information inside the WatchFace class

In this section, we will proceed to create our custom `WatchFace` class that will be used to invoke the methods within the watch face service and this class will be responsible for allocating and initializing the resources that our watch face requires.

First, we need to create a new class called `CustomWatchFace`, as follows:

1. From the **Project Navigator** window, expand the **wear** section and select and expand the **java** section.

2. Next, right-click and choose the **New | Java Class** menu option.

3. Then, enter `CustomWatchFace` to be used as the name for our class and click on the **OK** button to open the Android Studio code editor window.

   Our next step is to write the code that will communicate with our Android wearable device. For this, we will need to create a new class that will act as our watch service.

4. Open `CustomWatchFace.java` as shown in the preceding screenshot.

5. Next, enter the `import` statements as shown in the following code snippet:

```java
import android.content.Context;
import android.graphics.Canvas;
import android.graphics.Color;
import android.graphics.Paint;
import android.graphics.Rect;
import java.util.TimeZone;
import java.text.SimpleDateFormat;
import java.util.Calendar;
```

6. Now, modify the `CustomWatchFace` class as shown in the following code snippet:

```java
public class CustomWatchFace {
    private   final Paint   timeObject;
    private   final Paint   dateObject;
    private   final Paint   batteryObject;
    private         String  dateText;
    private         String  timeText;
    public          String  batteryText;

    private static final String TIME_FORMAT = "kk:mm:ss a";
    private static final String DATE_FORMAT = "EEE, dd MMM yyyy";

    // Declare our class constructor
    public static CustomWatchFace newInstance(Context context) {
        Paint batteryObject = new Paint();
            batteryObject.setColor(Color.RED);
            batteryObject.setTextSize(25);
            Paint timeObject = new Paint();
            timeObject.setColor(Color.GREEN);
            timeObject.setTextSize(35);
            Paint dateObject = new Paint();
```

```
            dateObject.setColor(Color.WHITE);
            dateObject.setTextSize(35);

            return new CustomWatchFace(timeObject, dateObject,
batteryObject);
        }
  CustomWatchFace(Paint objTime, Paint objDate, Paint
                objBattery) {
        this.timeObject = objTime;
        this.dateObject = objDate;
        this.batteryObject = objBattery;

        // Initialize our Battery Information
        batteryText = "Level: 0%";
        }
// Method to update the watch face each time an update
// has occurred
public void draw(Canvas canvas, Rect bounds) {
  canvas.drawColor(Color.BLACK);
  timeText = new
  SimpleDateFormat(TIME_FORMAT).format(Calendar.getInstance().
  getTime());
  dateText = new
  SimpleDateFormat(DATE_FORMAT).format(Calendar.getInstance().
  getTime());
  float timeXOffset = calculateXOffset(timeText, timeObject,
  bounds);
  float timeYOffset = calculateTimeYOffset(timeText,
  timeObject, bounds);
  canvas.drawText(timeText, timeXOffset, timeYOffset,
  timeObject);
  float dateXOffset = calculateXOffset(dateText, dateObject,
  bounds);
  float dateYOffset = calculateDateYOffset(dateText,
  dateObject);
  canvas.drawText(dateText, dateXOffset, timeYOffset +
  dateYOffset, dateObject);
  float batteryXOffset = calculateXOffset(batteryText,
  batteryObject, bounds);
```

```
      float batteryYOffset = calculateBatteryYOffset(batteryText,
      batteryObject);
canvas.drawText(batteryText, batteryXOffset, dateYOffset +
batteryYOffset, batteryObject);
      }
      // Calculate our X-Offset using our Time Label as the offset
      private float calculateXOffset(String text, Paint paint,
Rect watchBounds) {
float centerX = watchBounds.exactCenterX();
        float timeLength = paint.measureText(text);
        return centerX - (timeLength / 2.0f);
      }
      // Calculate our Time Y-Offset
      private float calculateTimeYOffset(String timeText, Paint
      timePaint, Rect watchBounds) {
        float centerY = watchBounds.exactCenterY();
        Rect textBounds = new Rect();
        timePaint.getTextBounds(timeText, 0,
        timeText.length(), textBounds);
        int textHeight = textBounds.height();
        return centerY + (textHeight / 2.0f);
      }
      // calculate our Date Label Y-Offset
      private float calculateDateYOffset(String dateText, Paint
      datePaint) {
        Rect textBounds = new Rect();
        datePaint.getTextBounds(dateText, 0,
        dateText.length(), textBounds);
        return textBounds.height() + 10.0f;
      }
      // Calculate our Battery Label Y-Offset
      private float calculateBatteryYOffset(String batteryText,
        Paint batteryPaint) {
        Rect textBounds = new Rect();
        batteryPaint.getTextBounds(batteryText, 0,
        batteryText.length(), textBounds);
        return textBounds.height() + 40.0f;
      }
```

```java
    public void setAntiAlias(boolean antiAlias) {
        batteryObject.setAntiAlias(antiAlias);
        timeObject.setAntiAlias(antiAlias);
        dateObject.setAntiAlias(antiAlias);
    }
    // Set each of our objects colors
    public void setColor(int red, int green, int white) {
        batteryObject.setColor(red);
        timeObject.setColor(green);
        dateObject.setColor(white);
    }
    // method to get our current timezone and update the time
field
    public void updateTimeZoneWith(String timeZone) {
        // Set our default time zone
        TimeZone.setDefault(TimeZone.getTimeZone(timeZone));
        // Get the current time for our current timezone
        timeText = new
        SimpleDateFormat(TIME_FORMAT).format(Calendar.
getInstance().
        getTime());
    }
}
```

In the preceding code snippet, we started by adding the `import` statements that are responsible for communicating. Also, we declared our `CustomWatchFace` class and the variables that will be responsible for holding our date, time, and battery data as well as the date and time formats defined by our `TIME_FORMAT` and `DATE_FORMAT` variables. Next, we declared our class constructor that will be called when the class is instantiated by our `WatchFaceService` class and will be responsible for setting the font sizes and colors for our `Date`, `Time`, and `Battery` objects.

The `draw` method is called whenever an update is required for the watch face canvas defined by our `WatchFaceService` class, which is handled by the `onTimeTick` callback that fires every 60 seconds by default. This also handles any UI updates when our watch face goes into ambient mode. We use the `Calendar` object declared in our `java.util.Calendar` package to obtain the current time and format using the `TIME_FORMAT` and `DATE_FORMAT` formatters.

Next, we make a call to each of the `calculateXOffset`, `calculateTimeYOffset`, `calculateDateYOffset`, and `calculateBatteryYOffset` methods to position each of our labels within our watch face using the watch face dimensions, and then make a call to our `drawText` method on our `WatchFace` canvas for each of our elements to add these to the watch face view.

The `setAntiAlias` method is used to minimize the pixels used in ambient mode that is being drawn within the watch face area so that the content is smooth. Our `WatchFaceService` class calls the `setColor` method whenever the view enters ambient or interactive mode and updates the colors accordingly. The `updateTimeZoneWith` method updates when the users adjust their time zone; the system broadcasts this event and the time will be automatically updated accordingly.

In our next section, we will create a custom Android watch face service class that will be used to communicate with our Google wear API, so that we can configure our system user interface.

Creating a custom watch face service class

In this section, we will proceed to create our custom watch face service class that will be used to invoke the methods within the watch face service API. This class will be responsible for allocating and initializing the resources that our watch face requires. Watch faces are services and shouldn't be confused with activities, as these types of services only accept touch input and voice commands as a form of interaction.

First, we need to create a new class called `WatchFaceService`:

1. From the **Project Navigator** window, expand the **wear** section and select and expand the **java** section.

2. Next, right-click and choose the **New | Java Class** menu option:

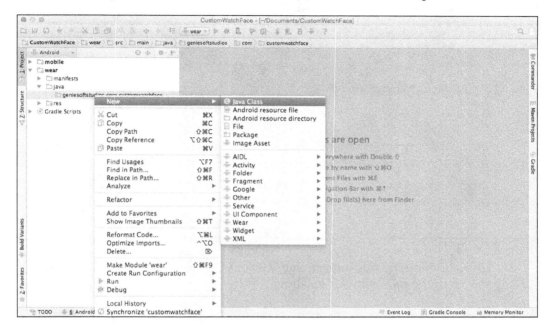

3. Then, enter `WatchFaceService` to be used as the name for our class and click on the **OK** button:

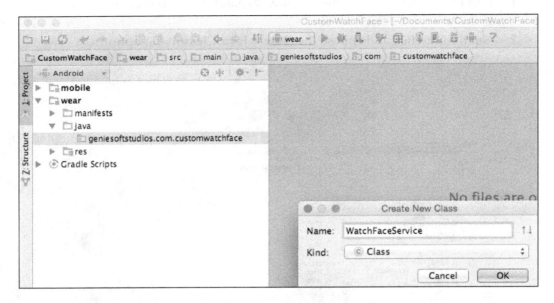

4. Upon clicking the **OK** button, the Android Studio code editor will open, as shown in the following screenshot:

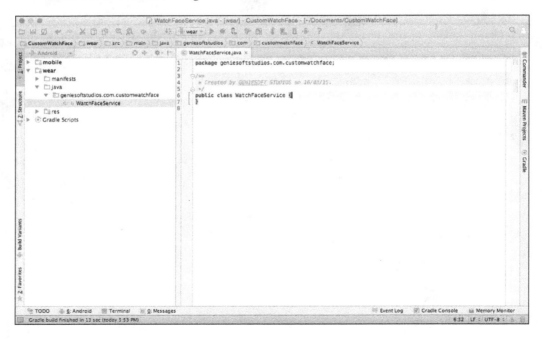

Our next step is to write the code that will communicate with our Android wearable device. For this, we will need to create a new class that will act as our watch service.

5. Open `WatchFaceService.java` as shown in the preceding screenshot.

6. Next, enter the `import` statements as shown in the following code snippet:

```
import android.content.BroadcastReceiver;
import android.content.Context;
import android.content.Intent;
import android.content.IntentFilter;
import android.graphics.Canvas;
import android.graphics.Color;
import android.graphics.Rect;
import android.os.BatteryManager;
import android.os.Handler;
import android.os.Looper;
import android.support.wearable.watchface.CanvasWatchFaceService;
import android.support.wearable.watchface.WatchFaceStyle;
import android.view.SurfaceHolder;
import java.util.Calendar;
```

7. Now, modify the `WatchFaceService` class as follows:

```
// Create our WatchFaceService Class
public class WatchFaceService extends CanvasWatchFaceService {
@Override
    public Engine onCreateEngine() {
        return new WatchFaceEngine();
    }
    // Create our WatchFaceEngine Class
    private class WatchFaceEngine extends
            CanvasWatchFaceService.Engine {
        }
......
}
```

8. In this step, we need to add the code for our `WatchFaceEngine` class by creating an `onCreate(SurfaceHolder holder)` method that will be called when the watch is instantiated and this will initialize the watch face, as shown in the following code snippet:

```
private CustomWatchFace watchFace;
private Handler clockTick;

@Override
        public void onCreate(SurfaceHolder holder) {
            super.onCreate(holder);
            setWatchFaceStyle(new
            WatchFaceStyle.Builder(WatchFaceService.this)
.setCardPeekMode(WatchFaceStyle.PEEK_MODE_SHORT)
.setAmbientPeekMode(WatchFaceStyle.
AMBIENT_PEEK_MODE_HIDDEN)
.setBackgroundVisibility(WatchFaceStyle.BACKGROUND_
            VISIBILITY_INTERRUPTIVE)
.setShowSystemUiTime(false)
.build());

 clockTick = new Handler(Looper.myLooper());
registerBatteryInfoReceiver();
startTimerIfNecessary();
watchFace =
CustomWatchFace.newInstance(WatchFaceService.this);
        }
```

9. Next, we need to create a method for the `WatchFaceEngine` class that will be called to update the watch display when the number of seconds has incremented. This can be done as follows:

```
    private void startTimerIfNecessary() {
      clockTick.removeCallbacks(timeRunnable);
      if (isVisible() && !isInAmbientMode()) {
        clockTick.post(timeRunnable);
      }
    }
```

10. Now, we need to create and implement a new class instance of a runnable object and include a method called `run` that starts executing the active class. The following code snippet shows how to do this:

```
private final Runnable timeRunnable = new Runnable() {
            @Override
            public void run() {
onSecondTick();
if (isVisible() && !isInAmbientMode()) {
                long TICK_PERIOD_MILLIS =
                Calendar.getInstance().get(Calendar.MILLISECOND);
                clockTick.postDelayed(this, TICK_PERIOD_MILLIS);
            }
        }
    };
    // Method to handle when the seconds are updating
    private void onSecondTick() {
        invalidateIfNecessary();
    }
    // stops any updates to the view, at which point the
  // onDraw method will be called at some point in the
  // future to refresh the view.
private void invalidateIfNecessary() {
if (isVisible() && !isInAmbientMode())
invalidate();
}
```

11. Here, we create the `onVisibilityChanged` method that is called when the watch face is visible and is responsible for registering the receiving methods when the time zone changes and starts the custom timer if the device is in interactive mode. When the watch face is not visible, this method stops the custom timer and unregisters the receiver for time zone changes. The `registerReceiver` and `unregisterReceiver` methods are implemented, for example, when the battery level changes. Have a look at the following code snippet:

```
@Override
        public void onVisibilityChanged(boolean visible) {
            super.onVisibilityChanged(visible);
            if (visible) {
registerTimeZoneReceiver();
```

```
                registerBatteryInfoReceiver();
                        }
                        else {
                            unregisterTimeZoneReceiver();
                            unregisterBatteryInfoReceiver();
                        }
            startTimerIfNecessary();
                }
            private void registerTimeZoneReceiver() {
                IntentFilter timeZoneFilter = new
    IntentFilter(Intent.ACTION_TIMEZONE_CHANGED);
                registerReceiver(timeZoneChangedReceiver,
                        timeZoneFilter);
            }
            private BroadcastReceiver timeZoneChangedReceiver = new
                BroadcastReceiver() {
                @Override
                public void onReceive(Context context, Intent intent){
                    if (Intent.ACTION_TIMEZONE_CHANGED.equals(intent.
    getAction())) {
    watchFace.updateTimeZoneWith(intent.getStringExtra("time-zone"));
                    }
                }
            };
    // method to unregister our detected timezone receiver
            private void unregisterTimeZoneReceiver() {
                unregisterReceiver(timeZoneChangedReceiver);
            }
            // Register a broadcast message to get the battery level
            private void registerBatteryInfoReceiver() {
                IntentFilter batteryInfoFilter = new
                        IntentFilter(Intent.ACTION_BATTERY_CHANGED);
                registerReceiver(batteryInfoChangedReceiver,
                batteryInfoFilter);
            }
            // Method to receive the message sent by the battery info
            // receiver class
            private BroadcastReceiver batteryInfoChangedReceiver = new
                    BroadcastReceiver() {
```

```
@Override
public void onReceive(Context context, Intent intent){
if
(Intent.ACTION_BATTERY_CHANGED.equals(intent.getAction()))
{
watchFace.batteryText =
String.valueOf("Battery: " +
intent.getIntExtra(BatteryManager.EXTRA_LEVEL, 0) + "%");
}
}
};
// method to unregister our battery information receiver
private void unregisterBatteryInfoReceiver() {
    unregisterReceiver(batteryInfoChangedReceiver);
}
```

12. In the following step, we create the `onDraw` method that will be responsible for handling and updating the watch face on the main runnable thread when the view comes out of the ambient mode. The `onTimeTick` method is called whenever the timer ticks and calls the `invalidate` method, which tells the system to call the `onDraw` method to redraw the watch face as shown in the following code snippet:

```
@Override
public void onDraw(Canvas canvas, Rect bounds) {
    super.onDraw(canvas, bounds);
    watchFace.draw(canvas, bounds);
}
@Override
public void onTimeTick() {
    super.onTimeTick();
    invalidate();
}
```

13. Next, we create the `onAmbientModeChanged` method that determines when the device changes between ambient and interactive modes. We perform a check to see if we are in ambient mode, and change the color of our display labels for `Date`, `Time`, and `Battery` labels before calling the `invalidate` method so that the system will redraw the watch face. This is done as follows:

```
@Override
public void onAmbientModeChanged(boolean inAmbientMode) {
    super.onAmbientModeChanged(inAmbientMode);
```

```
watchFace.setAntiAlias(!inAmbientMode);
if (!inAmbientMode) {
    watchFace.setColor(Color.RED, Color.GREEN,
                        Color.WHITE);
}
else watchFace.setColor(Color.GRAY, Color.GRAY,
                        Color.GRAY);
invalidate();
startTimerIfNecessary();
}
```

14. Now, we create the `onDestroy` method that handles the destroying of our callback methods declared by the register methods as follows:

```
@Override
public void onDestroy() {
    clockTick.removeCallbacks(timeRunnable);
    super.onDestroy();
}
}
```

In the preceding code snippets, we started by adding the import statements that will be responsible for allowing our application to communicate with the Android wearable. Then we added the code for our `WatchFaceEngine` class by creating `onCreate(SurfaceHolder holder)` that will be called when the watch is instantiated, as this will initialize the watch face.

In our next step, we create a `startTimerIfNecessary` method that will be called to update the watch display when the number of seconds has incremented and proceed to implement runnable class that includes a method called `run` that starts executing the active class. The `onVisibilityChanged` method is called when the watch face is visible, and is responsible for registering the receiving methods when the time zone changes, and starts the custom timer if the device is in interactive mode.

When the watch face is not visible, this method stops the custom timer and unregisters the receiver for time zone changes. The `registerReceiver` and `unregisterReceiver` methods are implemented for instance, when the battery level changes. The `onDraw` method is responsible for handling updates to the watch face on the main runnable thread when the view comes out of the ambient mode. The `onTimeTick` method is called whenever the timer ticks and calls the `invalidate` method, which tells the system to call the `onDraw` method to redraw the watch face.

We proceed to implement the `onAmbientModeChanged` method that will be responsible for determining whenever the device changes between the ambient and interactive modes. Next, we check to see if we are in ambient mode and change the color of our display labels for `Date`, `Time`, and `Battery` labels before calling the `invalidate` method, so that the system will redraw the watch face before finally destroying all of our callback methods declared by the register methods in the `onDestroy` method.

Debug your Android wearable app over Bluetooth

In this section, we will be taking a look at the steps involved in debugging your wearable application over Bluetooth. This process uses the Android debug bridge to facilitate communications between the handheld and the wearable devices by routing its debug output to the handheld device that is connected to your development machine.

To set up your device for debugging, follow these simple steps:

1. Enable USB debugging on the handheld device by opening the **Settings** app and then scrolling down till until you see **Developer options**.

2. From the **Developer options** section, scroll down to and enable **USB debugging** as shown in the following screenshot:

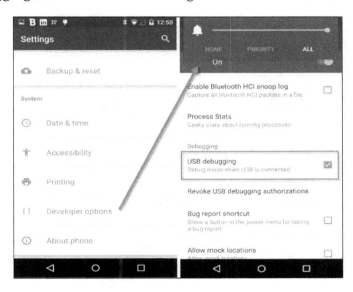

3. Next, open the Android Wear companion app on the handheld device.

4. Click on the **Settings** cog that is located on the top right-hand corner of the screen, scroll down, and enable **Debug over Bluetooth**, which is shown in the following screenshot:

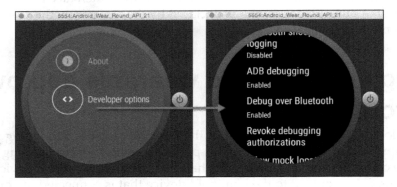

5. Next, connect your handheld device to your machine using USB and enter the following command line options, which are shown in the following screenshot:

```
$ ./adb forward tcp:4444 localabstract:/adb-hub
$ ./adb connect localhost:4444
```

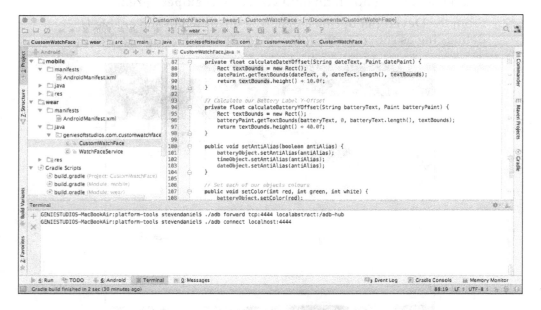

6. If you open the Android wearable app on your handheld device, you should see that the host and target statuses should have now changed to **connected**, as shown in the following screenshot:

Running apps directly on an Android Wear device

In this section, we will be taking a look at the steps involved in running an Android wearable app directly on the wearable device, instead of running this inside the Android simulator.

Before we can run the apps on our Android wearable device, first we need to make some changes to our project configuration as follows:

1. This can be achieved by choosing **Run | Edit Configurations...** from **Android Studio** menu as shown in the following screenshot:

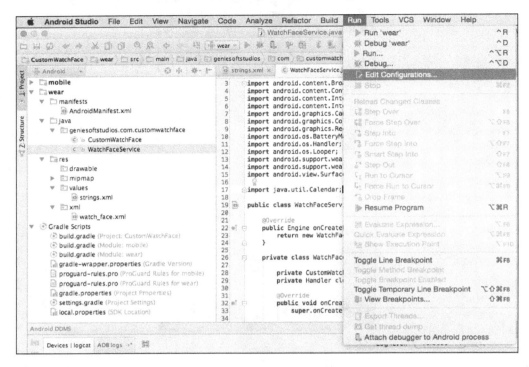

Next, since our project doesn't contain any activity modules, we need to make some adjustments to our project configuration.

2. From the **Run/Debug Configurations** screen, select the **wear** configuration option located within the **Android Application** section.

3. Then, select the **Do not launch Activity** option located within the **Activity** section.

4. Next, select the **Show chooser dialog** option located within the **Target Device** section, and click on the **OK** button to proceed to save your changes as shown in the following screenshot:

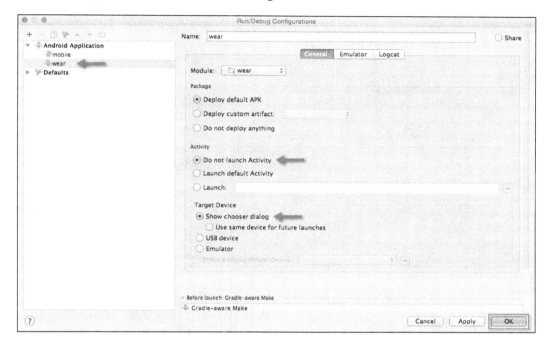

5. From the **Project Navigator** window, choose the **wear** section and then select the manifests folder.

6. Select the `AndroidManifest.xml` file as shown in the following screenshot:

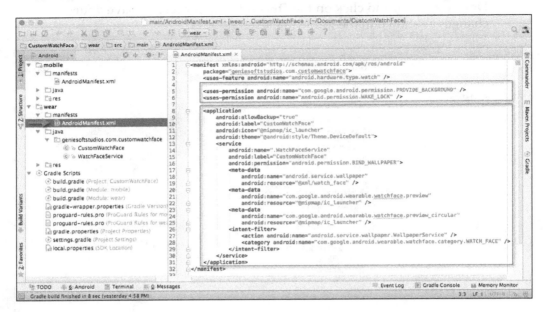

7. Next, under the `manifest` section of the wearable app we need to include permissions to allow our app to run within the wearable device. Enter the permissions as shown in the following code snippet:

```
<uses-permission android:name="com.google.android.permission.
PROVIDE_BACKGROUND"/>
<uses-permission android:name="android.permission.WAKE_LOCK"/>
```

8. Now modify the `<application>` section of the wearable app, as follows:

```
<application
        android:allowBackup="true"
        android:label="@string/app_name"
        android:icon="@mipmap/ic_launcher"
        android:theme="@android:style/Theme.DeviceDefault">
<service
android:name=".WatchFaceService"
android:label="@string/app_name"
android:permission="android.permission.
```

```
BIND_WALLPAPER">
<meta-data
android:name="android.service.wallpaper"
android:resource="@xml/watch_face" />
<meta-data
android:name="com.google.android.wearable.
watchface.preview"
android:resource="@mipmap/ic_launcher" />
<meta-data
android:name="com.google.android.wearable.
watchface.preview_circular"
android:resource="@mipmap/ic_launcher" />
<intent-filter>
<action
android:name="android.service.wallpaper.
                WallpaperService"/>
<category
android:name="com.google.android.wearable.
                watchface.category.WATCH_FACE"/>
</intent-filter>
</service>
</application>
```

9. Now, we can finally begin to compile, build, and run our application. Simply press *CMD + F9*, and choose your AVD or your Android wearable device from the list of available devices as shown in the following screenshot:

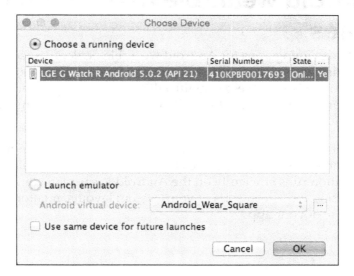

Once the wearable app has been installed on the Android wearable device, you should see our custom watch face being displayed, as shown in the following screenshot:

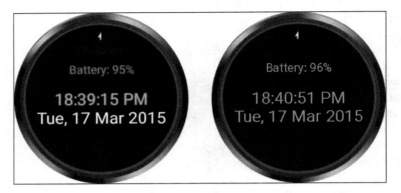

As you can see, creating Android watch faces is quite simple and you can create some really cool watch face designs. From the preceding screenshot, the image to the left shows the non-ambient version and the one to the right shows the image when the screen has dimmed, thus conserving battery life.

In our next section, we will take a look at the Android design principles, and the importance of designing user interfaces that are intuitive, consistent, and are designed with battery considerations in mind to conserve power.

The Android wearable user interface guidelines

The approach to designing apps for Android wearable devices, especially custom watch faces, needs to be substantially different from how you go about designing apps for phones or tablets, as these contain a different user experience. It is important to keep in mind and follow the Android UI design principles documentation that Google has provided.

This document describes the guidelines and principles that help you to design consistent user interfaces and experiences for your Android wearable apps, ensure that your application runs efficiently on the Android Wear platform, and it also involves considering the screen sizes of your custom layouts, memory limitations, and the ease of use of your app.

Other areas covered by the document are guidelines to ensure the consistency of your application as you navigate from screen to screen, as well as principles to design good user interfaces. This document also includes the following design guidelines that you will need to follow in your application, such as:

- Your application works for the round and square watch face designs, as well as their different screen resolutions.

- Your Android Wear app supports ambient mode when it is not being used, as this will conserve battery power. This involves dimming the screen and only using limited color, while keeping most of the pixels black. When the user taps on the screen to exit ambient mode, the screen can revert back to using full color and animations.

- Your app functionality allows the user to swipe down on the home screen to reveal the date and battery display, while providing the ability for further swiping down to turn off device sounds as well as preventing notifications from being displayed on the home screen.

There is also information relating to the proper use and appearance of system UI elements and controls for navigation, as well as creation of custom icons and images.

 To obtain further information about these guidelines, it is worthwhile to check the Android Wear Design Principles documentation at `https://developer.android.com/design/wear/principles.html`.

Packaging your Android wearable application

After you have finished testing your application to ensure that it is free of bugs, you are ready to publish your app to the world. Before this can happen, you must publish your wearable app directly inside a handheld application. This is due to the fact that your users cannot directly install your app to the wearable device.

Fortunately, this process is not that difficult, and in just a few steps you will be able to package your application using Android Studio. Let's get started by following these simple steps:

1. From the **Gradle Scripts** section of the **Project Navigator** window, select the **build.gradle (Module: mobile)** option as shown in the following screenshot:

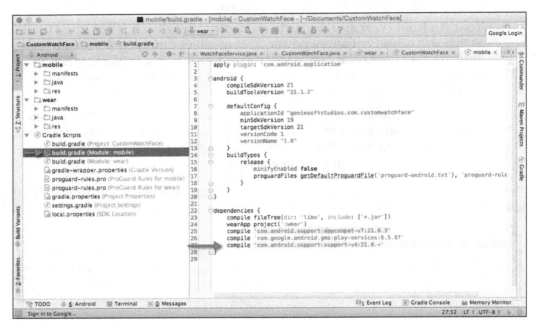

2. Next, under the `dependencies` section, enter the following code snippet:

```
compile 'com.android.support:support-v4:21.0.+'
```

 You need to ensure that both your wearable and handheld app modules contain the same package name and version, otherwise you will experience build errors.

3. From the `mobile` section of the **Project Navigator** window, select the `manifests` folder and then the `AndroidManifest.xml` file as follows:

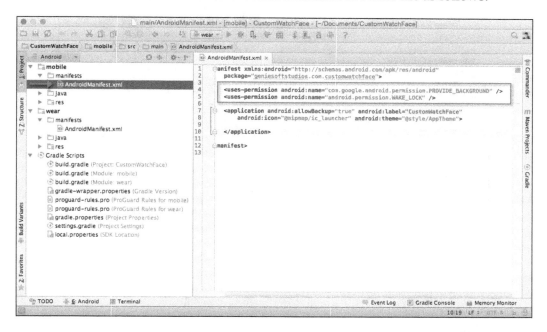

4. Next, under the `manifest` section of the handheld app, we need to include all permissions declared in our wearable app module in the manifest file of the handheld app module.

5. Then, enter the following permissions:

```
<uses-permission android:name="com.google.android.permission.
PROVIDE_BACKGROUND"/>
<uses-permission android:name="android.permission.WAKE_LOCK"/>
```

6. Now, navigate to **Build | Generate Signed APK...** as shown in the following screenshot:

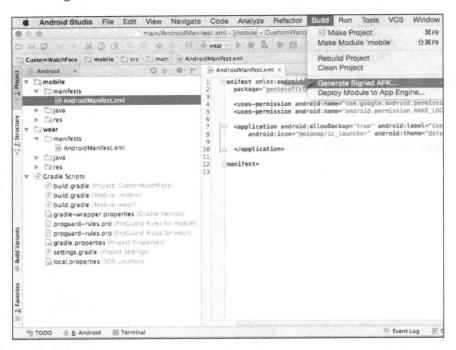

7. From the **Module** drop-down menu, choose the **mobile** option, and click on the **Next** button as shown in the following screenshot:

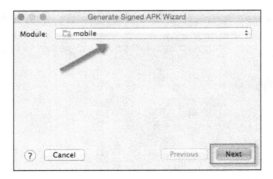

8. Next, from the **Generate Signed APKWizard** screen, click on the **Create new...** button, and then click on the **Next** button as follows:

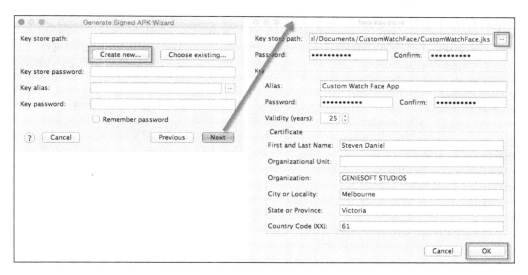

9. From the **New Key Store** screen, specify the location where you would like to store your key by clicking on the **...** button beside the **Key store path** text field.

 It is extremely important that you don't lose this key, otherwise you won't be able to upload new updates to the Google Play Store if the keys are different.

10. Then, provide a password to use for your key store file, an alias name to use for your application, populate the **Certificate** information, click on the **OK** button to dismiss the screen, and return back to the **Generate Signed APK Wizard** screen.

11. From the **Generate Signed APK Wizard** screen, click on the **Next** button, and then under the **Build Type** drop-down menu, choose the release option, and click on the **Finish** button.

12. Once the wizard has finished packaging your application, you will see the **Signed APK's generated successfully** dialog appear. Click on the **Reveal in Finder** button to show your `mobile-release.apk` packaged app as shown in the following screenshot:

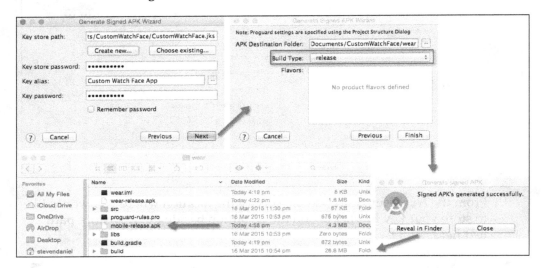

The preceding screenshot shows the workflow when generating a signed APK with the associated screens. As you can see, the APK will export your signed handheld application using Android Studio for your packaging process, and through this process, automatically include the wearable application portion embedded inside it.

 When a user downloads and installs your app, they will be installing the mobile component on their handheld device, which automatically pushes the wearable app to their smartwatch.

Summary

In this chapter, we learned how we can build a custom watch face by using Google's official API so that we can create our very own custom watch face service and custom watch face class to handle the presentation of a digital clock with date and time, as well as battery usage within the watch face layout.

Then we looked at how we can debug our wearable applications with Bluetooth that are running in the wearable device. Next, we spent some time learning about the design guidelines for Android wearable and the design considerations developers need to consider when designing their user interfaces, specifically to conserve battery power when running resource-intensive tasks.

In closing, we looked at the steps involved when packaging our wearable application so that it can be sent to users to be installed and used within the handheld mobile device.

In the next chapter, we will learn more about the data layer API and how we can synchronize data from the Android wearable with the handheld mobile device. We will learn about the message API that will provide us with the ability to send and receive messages, and finally we will learn how to build an Android Wear watch service to communicate with the data layer events.

Sending and Syncing Data

4

This chapter will provide you with the background and understanding of how you can effectively build applications that communicate between the Android handheld device and the Android wearable, send messages and compressed blob image data over Bluetooth to the Android wearable device, and then present this information within the wearable watch face area.

Android Wear comes with a number of APIs that will help to make communicating between the handheld and the wearable a breeze. We will be learning the differences between using `MessageAPI`, which is sometimes referred to as a "fire and forget" type of message, and `DataLayerAPI` that supports syncing of data between a handheld and a wearable, and `NodeAPI` that handles events related to each of the local and connected device nodes.

We will learn how to use `DataLayerAPI` to send an image from our handheld device to the wearable device, and then use `MessageAPI`, which will enable us to send and receive messages between the handheld and the wearable using the `DataAPI` event methods.

This chapter includes the following topics:

- Creating an Android wearable app to send and receive information
- Setting up the UI for our handheld app
- Setting up the UI for our Android wearable
- Establishing communication between the handheld and the wearable
- Sending and receiving messages between the handheld and the wearable device
- Sending and receiving images sent from the handheld to the wearable device

Creating a wearable send and receive application

In this section, we will take a look at how to create an Android wearable application that will send an image and a message, and display this on our wearable device. In the next sections, we will take a look at the steps required to send data to the Android wearable using `DataAPI`, `NodeAPI`, and `MessageAPIs`.

Firstly, create a new project in Android Studio by following these simple steps:

1. Launch Android Studio, and then click on the **File | New Project** menu option.

2. Next, enter `SendReceiveData` for the **Application name** field.

3. Then, provide the name for the **Company Domain** field.

4. Now, choose **Project location** and select where you would like to save your application code:

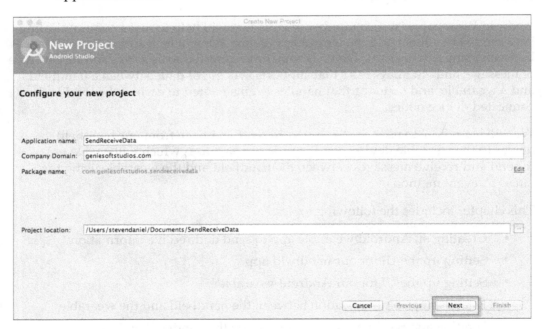

5. Click on the **Next** button to proceed to the next step.

   Next, we will need to specify the form factors for our phone/tablet and Android Wear devices using which our application will run. On this screen, we will need to choose the minimum SDK version for our phone/tablet and Android Wear.

6. Click on the **Phone and Tablet** option and choose **API 19: Android 4.4 (KitKat)** for **Minimum SDK**.

7. Click on the **Wear** option and choose **API 21: Android 5.0 (Lollipop)** for **Minimum SDK**:

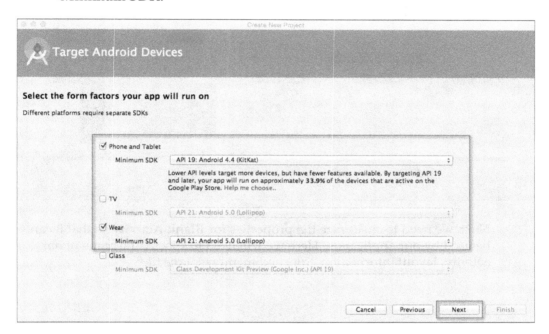

8. Click on the **Next** button to proceed to the next step.

   In our next step, we will need to add **Blank Activity** to our application project for the mobile section of our app.

9. From the **Add an activity to Mobile** screen, choose the **Add Blank Activity** option from the list of activities shown and click on the **Next** button to proceed to the next step:

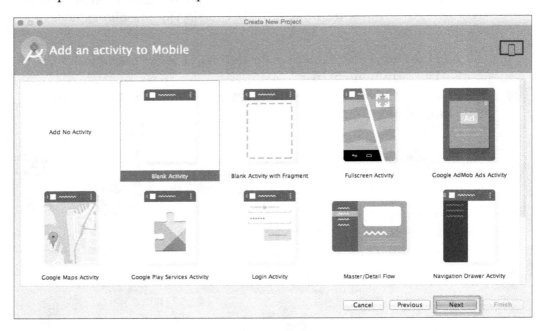

Next, we need to customize the properties for **Blank Activity** so that it can be used by our application. Here we will need to specify the name of our activity, layout information, title, and menu resource file.

10. From the **Customize the Activity** screen, enter `MobileActivity` for **Activity Name** shown and click on the **Next** button to proceed to the next step in the wizard:

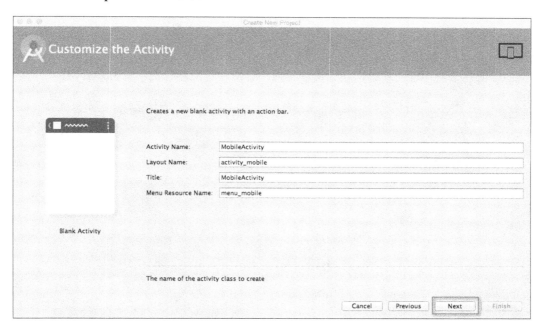

In the next step, we will need to add **Blank Activity** to our application project for the Android wearable section of our app.

11. From the **Add an activity to Wear** screen, choose the **Blank Wear Activity** option from the list of activities shown and click on the **Next** button to proceed to the next step:

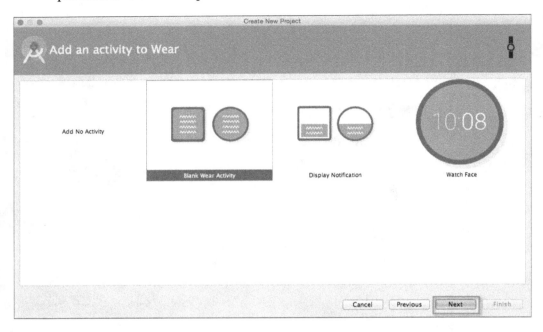

Next, we need to customize the properties for **Blank Wear Activity** so that our Android wearable can use it. Here we will need to specify the name of our activity and the layout information.

12. From the **Customize the Activity** screen, enter `WearActivity` for **Activity Name** shown and click on the **Next** button to proceed to the next step in the wizard:

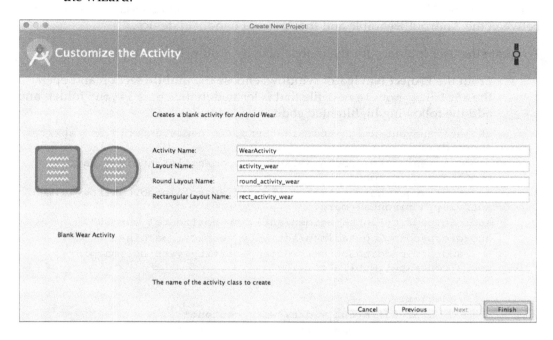

13. Finally, click on the **Finish** button and the wizard will generate your project and after a few moments, the Android Studio window will appear with your project displayed.

Creating a UI for the mobile activity

In this section, we need to build the UI for the Android handheld activity section of our application. This will enable us to communicate and send messages and images between the Android wearable and the Android handheld device.

To create the user interface for the mobile activity, follow these simple steps:

1. From the **Project Navigator** window, choose the **mobile** section and open the `activity_mobile.xml` file that is located in the `res | layout` folder, and add the following highlighted code:

```xml
<RelativeLayout xmlns:android="http://schemas.android.com/apk/res/
android"
xmlns:tools="http://schemas.android.com/tools" android:layout_
width="match_parent"
android:layout_height="match_parent" android:paddingLeft="@dimen/
activity_horizontal_margin"
android:paddingRight="@dimen/activity_horizontal_margin"
android:paddingTop="@dimen/activity_vertical_margin"
    android:paddingBottom="@dimen/activity_vertical_margin"
tools:context=".MobileActivity">

    <EditText
        android:layout_width="wrap_content"
        android:layout_height="wrap_content"
        android:inputType="textMultiLine"
        android:minEms="10"
        android:id="@+id/send_message_input"
      android:layout_marginTop="87dp"
        android:layout_alignParentTop="true"
        android:layout_centerHorizontal="true" />
    <Button
        android:layout_width="wrap_content"
        android:layout_height="wrap_content"
        android:text="@string/send_message_button"
        android:id="@+id/send_message_button"
        android:layout_below="@+id/send_message_input"
        android:layout_centerHorizontal="true" />
    <Button
      android:layout_width="wrap_content"
      android:layout_height="wrap_content"
      android:text="@string/send_image_button"
      android:id="@+id/send_image_button"
```

```
        android:layout_below="@+id/send_message_button"
        android:layout_centerHorizontal="true" />
</RelativeLayout>
```

2. Again from the **Project Navigator** window, choose the **mobile** section, open the `strings.xml` file that is located in the `res | values` folder, and add the highlighted code as follows:

```
<resources>
    <string name="app_name">SendReceiveData</string>
    <string name="action_settings">Settings</string>
    <string name="send_image_button">Send an image to
    Android Wear</string>
    <string name="send_message_button">Send the
    Message to Android Wear</string>
    <string name="send_message_text">Enter a message
    to send</string>
</resources>
```

3. In the same **Project Navigator** window, choose the **mobile** section, select the `manifests` folder, and then select the `AndroidManifest.xml` file as shown in the following screenshot:

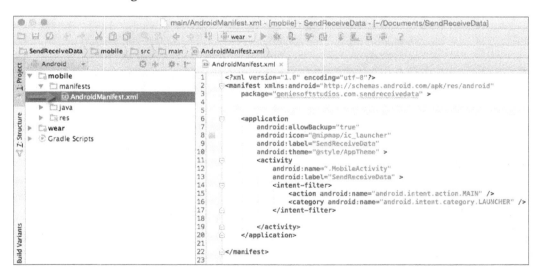

4. Next, under the `manifest` section of the mobile handheld app, we need to include permissions to allow our app to run within the handheld device. Enter the following permission:

```
<uses-permission android:name="android.permission.INTERNET"></uses-permission>
```

5. Now, modify the `<application>` section of the handheld app and enter the code sections highlighted in the following snippet:

```
<application
    android:allowBackup="true"
    android:icon="@mipmap/ic_launcher"
    android:label="@string/app_name"
    android:theme="@style/AppTheme" >
    <activity
        android:name=".MobileActivity"
        android:label="@string/app_name" >
        <intent-filter>
        <action
        android:name="android.intent.action.MAIN" />
        <category
        android:name="android.intent.category.LAUNCHER" />
        </intent-filter>
            <meta-data
            android:name="com.google.android.gms.version"
            android:value="@integer/google_play_services_version"/>
    </activity>
</application>
```

In our next section, we will need to create the user interface for the wearable portion of our application. This will be used to display messages and images sent from the handheld device to the wearable watch area.

Creating a UI for the wear activity

In this section, we need to build the user interface for the Android wearable activity section of our application. This will enable us to receive messages and images sent from the Android handheld device and have this information presented within the Android wearable watch area.

Perform the following steps:

1. From the **Project Navigator** window, open the `rect_activity_wear.xml` file that is located in the `res | layout` folder, and add the following code:

```xml
<?xml version="1.0" encoding="utf-8"?>
<RelativeLayout
    xmlns:android="http://schemas.android.com/apk/res/android"
    xmlns:tools="http://schemas.android.com/tools"
    android:layout_width="match_parent"
    android:layout_height="match_parent"
    android:orientation="vertical"
    tools:context=".WearActivity"
    tools:deviceIds="wear_square">

    <TextView
        android:layout_width="92dp"
        android:layout_height="31dp"
        android:minEms="10"
        android:id="@+id/received_message_input"
        android:layout_marginTop="20dp"
        android:layout_alignParentTop="true"
        android:layout_centerHorizontal="true"
        android:layout_gravity="center_horizontal" />

    <ImageView
        android:layout_width="111dp"
        android:layout_height="95dp"
        android:id="@+id/received_image_input"
        android:layout_below="@+id/received_message_input"
        android:adjustViewBounds="false"
        android:layout_marginLeft="10dp"
        android:layout_marginTop="20dp"
        android:layout_marginRight="10dp"
        android:layout_marginBottom="100dp"
        android:layout_gravity="center_horizontal" />
</RelativeLayout>
```

2. Next, open the `round_activity_wear.xml` file that is located in the `res | values` folder within **Project Navigator** and add the code sections as follows:

```xml
<?xml version="1.0" encoding="utf-8"?>
<RelativeLayout  xmlns:android="http://schemas.android.com/apk/
res/android"
xmlns:tools="http://schemas.android.com/tools" android:layout_
width="match_parent"
android:layout_height="match_parent" tools:context=".WearActivity"
tools:deviceIds="wear_round">
```

```
<TextView
    android:layout_width="wrap_content"
    android:layout_height="wrap_content"
    android:minEms="10"
    android:id="@+id/received_message_input"
    android:layout_marginTop="87dp"
    android:layout_alignParentTop="true"
    android:layout_centerHorizontal="true" />
<ImageView
    android:layout_width="match_parent"
    android:layout_height="match_parent"
    android:id="@+id/received_image_input"
    android:layout_below="@+id/received_message_input"/>

</RelativeLayout>
```

3. Now from the **Project Navigator** window, choose the **wear** section, select the `manifests` folder, and then select the `AndroidManifest.xml` file as shown in the following screenshot:

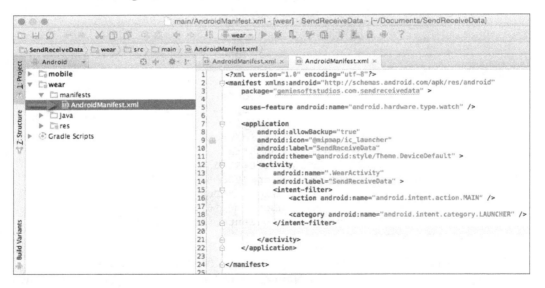

4. Then, modify the `<application>` section of the wearable app, and enter the following highlighted code sections:

```
<application
    android:allowBackup="true"
    android:icon="@mipmap/ic_launcher"
```

```
        android:label="@string/app_name"
        android:theme="@android:style/Theme.DeviceDefault" >
        <activity
            android:name=".WearActivity"
            android:label="@string/app_name" >
        <intent-filter>
          <action
          android:name="android.intent.action.MAIN" />
          <category
          android:name="android.intent.category.LAUNCHER" />
        </intent-filter>
        <!-- Allow communication with Google Play
            Services -->
        <meta-data
        android:name="com.google.android.gms.version"
          android:value="@integer/
          google_play_services_version"/>
        </activity>
    </application>
```

5. Now, from the **Project Navigator** window open the `MobileActivity.java` file as shown in the following screenshot:

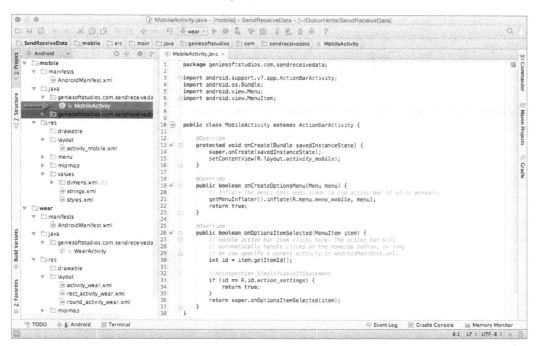

6. Then, add the following `import` statements to the `MobileActivity.java` file with the highlighted entries:

```
import android.support.v7.app.ActionBarActivity;
import android.os.Bundle;
import android.view.Menu;
import android.view.MenuItem;
import android.util.Log;
import android.view.View;
import java.io.ByteArrayOutputStream;
import java.net.URL;
import java.util.Random;
import com.google.android.gms.common.ConnectionResult;
import com.google.android.gms.common.api.GoogleApiClient;
import com.google.android.gms.wearable.Node;
import com.google.android.gms.wearable.NodeApi;
import com.google.android.gms.wearable.PutDataMapRequest;
import com.google.android.gms.wearable.Wearable;
import com.google.android.gms.common.api.GoogleApiClient.
ConnectionCallbacks;
import com.google.android.gms.common.api.GoogleApiClient.
OnConnectionFailedListener;
import com.google.android.gms.wearable.Asset;
import com.google.android.gms.wearable.DataMap;
import android.graphics.Bitmap;
import android.graphics.BitmapFactory;
import android.widget.Button;
import android.widget.EditText;
```

In the preceding code snippet, we started by adding the import statements that will be responsible for handling communication and will enable us to send messages and image data between our Android handheld device and the Android wearable. In the next section, we will be taking a look at how we can establish connections between our handheld and mobile device, as well as hook up buttons and text field controls to their associated method events.

Establishing connections for the mobile activity

Until now, we have created our project and the user interface for the mobile in our application. Also, we have set up our permissions so that our application can communicate with the Internet and Google Play Services.

In order for our app to communicate between the handheld and the Android wearable, we will need to establish a connection:

1. From the **Project Navigator** window, expand the **mobile** section, select, and expand the **java** section.

2. Next, double-click to open `MobileActivity.java`, and add the following code snippet:

```java
public class MobileActivity extends ActionBarActivity {

    private GoogleApiClient mGoogleApiClient;
    private static final String LOG_TAG = "MobileActivity";

    // establishes a connection between the mobile and wearable
    private void initGoogleApiClient() {
        if (mGoogleApiClient != null &&
        mGoogleApiClient.isConnected()) {
            Log.d(LOG_TAG, "Connected");
        }
        else
        {
            // Creates a new GoogleApiClient object with all
            // connection callbacks
            mGoogleApiClient = new GoogleApiClient.Builder(this)
            .addConnectionCallbacks(new ConnectionCallbacks() {
                @Override
                public void onConnected(Bundle connectionHint) {
                    Log.d(LOG_TAG, "onConnected: " + connectionHint);
                }
                @Override
                public void onConnectionSuspended(int
                cause) {
```

```
            Log.d(LOG_TAG, "onConnectionSuspended: "
            + cause);
        }
    })
    .addOnConnectionFailedListener(new
      OnConnectionFailedListener() {
      @Override
      public void onConnectionFailed(ConnectionResult
        result) {
          Log.d(LOG_TAG, "onConnectionFailed: " + result);
      }
    })
    .addApi(Wearable.API)
    .build();

    // Make the connection
    mGoogleApiClient.connect();
    }
}
```

3. Now, we need to create the activity callback methods that will be responsible for starting, stopping, resuming, and terminating the app, as shown in the following code snippet:

```
@Override
  protected void onStart() {
      super.onStart();
      initGoogleApiClient();
  }

@Override
protected void onStop() {
    super.onStop();
    mGoogleApiClient.disconnect();
}

  @Override
  protected void onResume() {
      super.onResume();
      initGoogleApiClient();
  }
```

```
@Override
protected void onDestroy() {
    super.onDestroy();
    mGoogleApiClient.disconnect();
}
```

In the preceding code snippet, we started by declaring our `GoogleApiClient` object variable that will be responsible for establishing and handling the connection between the Android handheld and the Android wearable device.

Next, we created our `initGoogleApiClient()` method that implements the `GoogleApiClient.ConnectionCallbacks` interface to handle all of the connection callbacks returned by `GoogleApiClient`. When `GoogleApiClient` has successfully established a connection with the wearable device, the callback method calls the `onConnected` method. If any errors have been detected, these will be caught by the `onConnectionFailed` callback method.

Then, we declare our `onStart()`, `onStop()`, `onResume()`, and `onDestroy()` activity class methods. The `onStart()` method will be called whenever the activity becomes visible and is displayed to the user, and the `onResume()` method is called after the `onStart()` method when the activity is displayed in the foreground. The `onStop()` method is called when the activity is no longer visible to the user, which happens when another activity has been resumed, or the current one is being destroyed. The `onDestroy()` method will be called once the activity has been removed from the activity chain, and is responsible for destroying any memory that has been previously allocated to variables from the memory.

Sending messages to the Android wearable

In this section, we will be taking a look at the steps involved in sending a message to your Android wearable device. This process is quite simple, and our next step is to write the code that will communicate between our Android wearable and the handheld device:

1. From the **Project Navigator** window, open the `MobileActivity.java` file.

2. Next, modify the `onCreate(Bundle savedInstanceState)` method and enter the code highlighted as follows:

```
@Override
protected void onCreate(Bundle savedInstanceState) {
```

```
  super.onCreate(savedInstanceState);
  setContentView(R.layout.activity_mobile);

  // Get a pointer to our buttons and textField
  final Button mSendMessageButton = (Button)
  findViewById(R.id.send_message_button);
  final EditText mSendMessageInput = (EditText)
  findViewById(R.id.send_message_input);

  // Set up our hint message for our Text Field
  mSendMessageInput.setHint(R.string.send_message_text);
```

3. Next, we need to set up an `onClickListener` method in `sendMessageButton` as shown in the following code snippet:

```
// Set up our send message button onClick method handler
mSendMessageButton.setOnClickListener(new
View.OnClickListener() {
  @Override
  public void onClick(View v) {
    // Create a new thread to send the entered message
    Thread thread = new Thread(new Runnable()
    {
      @Override
      public void run()
      {
        try {
          String messageText =
            mSendMessageInput.getText().toString();
          NodeApi.GetConnectedNodesResult nodes =
            Wearable.NodeApi.
          getConnectedNodes(mGoogleApiClient).await();
          for (Node node : nodes.getNodes()) {
            result = Wearable.MessageApi.
            sendMessage(mGoogleApiClient,
            node.getId(), "/message",
            messageText.getBytes() ).await();
          }
          runOnUiThread(new Runnable() {
            @Override
            public void run() {
```

```
                mSendMessageInput.getText().clear();
              }
            });
          }
          catch (Exception e) {
            Log.e(LOG_TAG,
            e.getMessage());
          }
        }
      });
      // Starts our Thread
      thread.start();
      Log.d(LOG_TAG, "Message has been sent");
    }
  });
```

In the preceding code snippet, we started by declaring two objects,
`mSendMessageButton` and `mSendMessageInput`. These objects contain the reference
to our button and the `EditTextField` will hold the text entered by the user. We
proceed to set some properties for each of these fields. Next, we proceed to set up
a `setOnClickListener` object that will handle and respond to the events when the
user has tapped on the **Send the Message to Android Wear** button.

In our next step, we create a new thread object that inherits from the `Thread`
class that includes a `run` method that will be used to execute the active class and
then declare a `messageText` string variable. This extracts the text entered by the
user, and then we use `NodeApi.getConnectedNodes()` to get a list of all nodes that
are currently connected to the device. In most cases, the `getConnectedNodes()`
method will return back a single node, but to make this future proof we iterate
over all the connected nodes to handle cases where the user may be signed in to
multiple devices.

Once we have the list, we send a message to each of the nodes using `Wearable.`
`MessageApi.sendMessage` that makes reference to `GoogleApiClient`, the current
node ID, the path used to determine the type of message being sent, and finally the
message payload, which is defined as a byte array. We use the await property to
block our wearable UI until the task completes.

Once the message has been sent, we clear our `mSendMessageInput` `EditText` field
to allow the user to continue to enter additional text. Any errors encountered will
be caught in the `catch (Exception e)` block. Then we call the start method of our
thread and display a message to our **Log** window to denote that the message has
been sent successfully.

Receiving messages using MessageAPI

In our previous section, we looked at how we can use `MessageAPI` to send messages to the Android wearable. In this section, we will take a look at how we can retrieve this message and display it on our Android wearable device:

1. From the **Project Navigator** window, open the `WearActivity.java` file:

2. Next, add the following highlighted `import` statements in the `WearActivity.java` file:

```java
import android.app.Activity;
import android.os.Bundle;
import android.support.wearable.view.WatchViewStub;
import android.widget.TextView;
import android.graphics.Bitmap;
import android.graphics.BitmapFactory;
import android.os.Handler;
import android.widget.ImageView;
import android.util.Log;
import com.google.android.gms.common.ConnectionResult;
```

```
import com.google.android.gms.common.api.GoogleApiClient;
import com.google.android.gms.wearable.Asset;
import com.google.android.gms.wearable.DataApi;
import com.google.android.gms.wearable.DataEvent;
import com.google.android.gms.wearable.DataEventBuffer;
import com.google.android.gms.wearable.DataMapItem;
import com.google.android.gms.wearable.MessageApi;
import com.google.android.gms.wearable.MessageEvent;
import com.google.android.gms.wearable.Wearable;
import java.io.InputStream;
```

3. Now, modify the `WearActivity` class as shown in the following code:

```
public class WearActivity extends Activity {

    private TextView mTextView;
    private GoogleApiClient mGoogleApiClient;
    private static final String LOG_TAG =  "WearActivity";

    @Override
    protected void onCreate(Bundle savedInstanceState) {
        super.onCreate(savedInstanceState);
        setContentView(R.layout.activity_wear);
        final WatchViewStub stub = (WatchViewStub)
        findViewById(R.id.watch_view_stub);
        stub.setOnLayoutInflatedListener(new
        WatchViewStub.OnLayoutInflatedListener() {
          @Override
          public void onLayoutInflated(WatchViewStub
          stub) {
            mTextView = (TextView)
            stub.findViewById(R.id.received_message_input);
          }
        });

                // Establish our connection
        initGoogleApiClient();
    }
```

4. Here, we need to create a new `initGoogleApiClient` method and add the code that will be called when the watch is instantiated. This will initialize the watch area, as follows:

```
// establishes a connection between the mobile and wearable
private void initGoogleApiClient()
{
  mGoogleApiClient = new GoogleApiClient.Builder(this)
  .addConnectionCallbacks(new
  GoogleApiClient.ConnectionCallbacks() {
    @Override
    public void onConnected(Bundle connectionHint) {
      Log.d(LOG_TAG, "onConnected: " +
      connectionHint);
      Wearable.MessageApi.addListener(
      mGoogleApiClient, messageListener);
    }
    @Override
    public void onConnectionSuspended(int cause) {
      Log.d(LOG_TAG, "onConnectionSuspended:
      " + cause);
    }
  })
  .addOnConnectionFailedListener(new
  GoogleApiClient.OnConnectionFailedListener() {
    @Override
    public void onConnectionFailed(ConnectionResult
    result) {
      Log.d(LOG_TAG, "onConnectionFailed: "
      + result);
    }
  })
  .addApi(Wearable.API)
  .build();

  mGoogleApiClient.connect();
}
```

5. Next, we need to create an `onMessageReceived` new method and add the following code that will be called when the wearable device receives the message from the handheld device:

```
MessageApi.MessageListener messageListener = new MessageApi.
MessageListener() {
        @Override
        public void onMessageReceived(final MessageEvent
            messageEvent) {
            runOnUiThread(new Runnable() {
                @Override
                public void run() {
                    if (messageEvent.getPath().
                        equalsIgnoreCase("/message")) {
                        Log.i(LOG_TAG, new
                        String(messageEvent.getData()));
                        mTextView.setText(new
                        String(messageEvent.getData()));
                    }
                }
            });
        }
};
```

6. Now, we need to create the `onStop` method that will handle clean up and destruction of our `GoogleApiClient` connections:

```
@Override
protected void onStop() {
  super.onStop();
  mGoogleApiClient.disconnect();
}
```

7. Next, we need to create the `onDestroy` method that will handle clean up and destruction of our `GoogleApiClient` connections.

```
@Override
protected void onDestroy() {
    super.onDestroy();
    mGoogleApiClient.disconnect();
}
```

8. In this step, we need to install the app on both Android handheld device and Android wearable device so that messages can be communicated between them. Choose the mobile configuration from the drop-down menu as shown in the following screenshot:

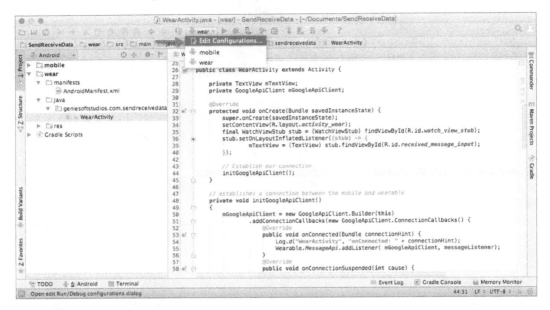

9. Then, install the app on our Android wearable device by following the same steps that we did for our Android handheld device, but choose the wear configuration from the drop-down menu as shown in the preceding screenshot.

In the preceding code snippet, we start by getting a reference to our `mTextView` object that will be used to display the message received from the Android handheld device. Next, we make a call to our `initGoogleApiClient` method that is responsible for establishing connection between our Android handheld and wearable device. Once the `GoogleApiClient` object has established a successful connection, the `onConnected` method is called where we set up a listener service on our `MessageApi` that will capture any messages our wearable class receives, and these will be handled by the `onMessageReceived` method.

In our next step, we create a new `messageListener` object that inherits from the `MessageApi.MessageListener` class that includes a `onMessageReceived` method. This method will be responsible for receiving messages sent from the Android handheld device and accepts a `messageEvent` variable that contains the message data object.

The `runOnUiThread` method is called only when a message data object has been received. Then we use `getPath` to get the path on which the message is being delivered and compare this to the path that we declared within our mobile activity class.

Next, we use the `getData` method to get the data passed by the message and then display this message on our **Log** window to denote that the message has been received successfully. This is done before assigning the contents of the message to our `mTextView` object, so that this can be displayed within the watch area. Next, we create the `onStop()` method that will be called when the activity is no longer visible to the user, or the `GoogleApiClient` variable has been destroyed by the `onDestroy` method.

Finally, we can begin to compile, build, and run our application. Simply press *CMD + F9* and choose your AVD or Android wearable device from the list of available devices as shown in the following screenshot:

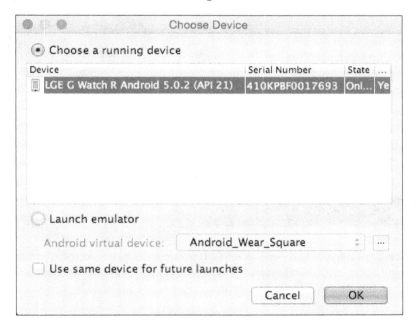

Once the wearable app has been installed on the Android handheld and Android wearable devices, you should see your entered message appear within the watch area, as follows:

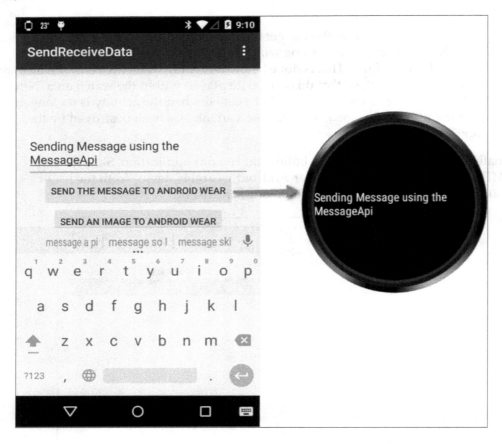

As you can see, using `MessageApi` to send messages is quite simple. In our next section, we will take a look at how we can use `DataApi` to send an image from the handheld device to the Android wearable.

Transferring image data to the Android wearable

In our previous section, we looked at how we can send messages from our Android handheld device and display this information within our wearable watch area. In this section, we will take a look at the steps involved in downloading an image from the Internet, compressing the file contents in memory before sending this to the Android wearable, and displaying this within the watch area:

1. From the **Project Navigator** window, open the `MobileActivity.java` file.

2. Next, modify the `onCreate(Bundle savedInstanceState)` method and enter the following highlighted code sections:

```
@Override
protected void onCreate(Bundle savedInstanceState) {
    super.onCreate(savedInstanceState);
    setContentView(R.layout.activity_mobile);

    // Get a pointer to our buttons and textField
    final Button mSendMessageButton = (Button)
    findViewById(R.id.send_message_button);
    final Button mSendImageButton = (Button)
    findViewById(R.id.send_image_button);
    final EditText mSendMessageInput = (EditText)
    findViewById(R.id.send_message_input);

    // Set up our hint message for our Text Field
    mSendMessageInput.setHint(R.string.send_message_text);
```

3. Now, we need to set up an `onClickListener` method in our `sendImageButton` as shown in the following code snippet:

```
// Set up our send image button onClick method handler
mSendImageButton.setOnClickListener(new
    View.OnClickListener() {
    @Override
    public void onClick(View v) {
        // Create a new thread to send the downloaded image
        Thread thread = new Thread(new Runnable(){
        @Override
```

```
        public void run()
        {
          // Declare our image variable to hold the URL
          String imageName =
          "http://www.androidcentral.com/sites/"
          + "androidcentral.com/files/styles/w550h500/public/" +
          "wallpapers/batdroid-blj.jpg";
          try {
            PutDataMapRequest request =
            PutDataMapRequest.create("/image");
            DataMap map = request.getDataMap();
            URL url = new URL(imageName);
            Bitmap bmp =
              BitmapFactory.decodeStream(url.openConnection().
            getInputStream());
            final ByteArrayOutputStream byteStream = new
            ByteArrayOutputStream();
            bmp.compress(Bitmap.CompressFormat.PNG, 100,
            byteStream);

            // Creates an image asset from the chosen image
            Asset asset =
            Asset.createFromBytes(byteStream.toByteArray());
            Random randomGenerator = new Random();
            int randomInt = randomGenerator.nextInt(1000);
            map.putInt("Integer", randomInt);
            map.putAsset("androidImage", asset);
            Wearable.DataApi.putDataItem(mGoogleApiClient,
            request.asPutDataRequest());
          }
          catch (Exception e) {
            Log.e(LOG_TAG, e.getMessage());
          }
      });
      // Starts our Thread
      thread.start();
      Log.d(LOG_TAG, "Image has been sent");
    }
  });
}
```

In the preceding code snippet, we started by declaring our `mSendImageButton` object variable that will be responsible for downloading the image from the Internet and sending this to the Android wearable device. Next, we proceed to set up a `setOnClickListener` object that will handle and respond to the events when the user has tapped on the **Send an Image to Android Wear** button. In our next step, we create a new thread object that inherits from the `Thread` class that includes a `run` method that will be used to execute the active class. Then we declare an `imageName` string variable that will be used to hold the path for our image that will be downloaded.

Next, we create an object variable request that references the `PutDataMapRequest` object that will be used to hold our image data. Then we call the `getDataMap` method to our request object to get the data map, so that we can proceed to assign values to this object that will be stored in the map object variable. In our next step, we declare a `url` object that uses the URL class that converts the `imageName` string into a URL object. Then we use `BitmapFactory.decodeStream` to connect to the Internet, return the object data, and assign the contents to our `bmp` object variable, then declare our `byteStream` object that creates `ByteArrayOutputStream` with a default size of `32` bytes.

 If `byteStream` sends more than `32` bytes, the underlying byte array will automatically expand.

In our next step, we write a compressed version of the bitmap to the specified bitmap stream, compress for maximum quality, and then pass our `byteStream` output stream to write the compressed data.

 The `path` string is a unique identifier for the data item that allows you to access it from either side of the connection. The path must begin with a forward slash.

Next, we need to create an `asset` image asset object variable from the compressed image data and convert the `byteStream` data into a byte array. **Assets** are objects that are used to send binary blobs of data, such as images. When you send **blob** data to the Android wearable, the system will automatically take care of the transfer for you, and conserve bandwidth by caching large assets to avoid retransmission.

The term blob is essentially a collection of binary data that is stored as a single entity and can be images, audio, or other multimedia objects.

Then we instantiate our `Random` class that constructs a random number generator with an initial state that is unlikely to be duplicated by a subsequent instantiation, and then assign this to our `randomGenerator` variable.

After this, declare a `randomInt` variable and generate a random number from one to one thousand. Then we create an `Integer` data item, and assign the random value to this. This is to ensure that we can send the image multiple times to the Android wearable device. Next, we proceed to do the same for our `androidImage` data item, and assign the created image asset data. Finally, we call the `DataApi.putDataItem` method, which adds the image data to the Android wearable network and sends the data to the Android wearable.

Receiving image data using DataApi

In the previous section, we looked at how we can use the `DataApi.putDataItem` method to send a binary image data to the Android wearable. In this section, we will be taking a look at how we can retrieve the image and display it within our Android wearable watch area:

1. From the **Project Navigator** window, open the `WearActivity.java` file.

2. Next, modify the `WearActivity` class and add the following code snippet:

```
public class WearActivity extends Activity {
    private TextView mTextView;
    private ImageView imageView;
    private GoogleApiClient mGoogleApiClient;
    private Bitmap imageBitmap;
    private final Handler imageHandler = new
    Handler();

    @Override
    protected void onCreate(Bundle savedInstanceState) {
        super.onCreate(savedInstanceState);
        setContentView(R.layout.activity_wear);
        final WatchViewStub stub = (WatchViewStub)
```

```
findViewById(R.id.watch_view_stub);
stub.setOnLayoutInflatedListener(new
WatchViewStub.OnLayoutInflatedListener() {
  @Override
  public void onLayoutInflated(WatchViewStub
  stub) {
  mTextView = (TextView)
    stub.findViewById(R.id.received_message_input);
  imageView = (ImageView)
    stub.findViewById(R.id.received_image_input);
  }
});

        // Establish our connection
initGoogleApiClient();
}
```

3. Then, we modify the `initGoogleApiClient` method and add the following highlighted code:

```
// establishes a connection between the mobile and wearable
private void initGoogleApiClient()
{
  mGoogleApiClient = new GoogleApiClient.Builder(this)
  .addConnectionCallbacks(new
  GoogleApiClient.ConnectionCallbacks() {
    @Override
    public void onConnected(Bundle connectionHint) {
      Log.d(LOG_TAG, "onConnected: " +
      connectionHint);
      Wearable.DataApi.addListener(
      mGoogleApiClient, onDataChangedListener);
      Wearable.MessageApi.addListener(
      mGoogleApiClient, messageListener);
    }
    @Override
    public void onConnectionSuspended(int cause) {
      Log.d(LOG_TAG, "onConnectionSuspended:
      " + cause);
```

```
      }
    })
    .addOnConnectionFailedListener(new
  GoogleApiClient.OnConnectionFailedListener() {
      @Override
      public void onConnectionFailed(ConnectionResult
      result) {
        Log.d(LOG_TAG, "onConnectionFailed: "
        + result);
      }
    })
    .addApi(Wearable.API)
    .build();

    mGoogleApiClient.connect();
}
```

4. Next, we create a onDataChanged(DataEventBuffer dataEvents) listener method and add the following code that will be called when the wearable device receives the image from the handheld device:

```
public DataApi.DataListener onDataChangedListener = new DataApi.
DataListener() {
  @Override
  public void onDataChanged(DataEventBuffer dataEvents){
    for (DataEvent event : dataEvents) {
      if (event.getType() ==
      DataEvent.TYPE_CHANGED &&
      event.getDataItem().getUri().getPath()
      .equals("/image")) {
        DataMapItem dataMapItem =
        DataMapItem.fromDataItem(
        event.getDataItem());
        Asset imageAsset =
        dataMapItem.getDataMap().
        getAsset("androidImage");

        imageBitmap =
        loadBitmapFromAsset(imageAsset);
        // Process our received image bitmap
```

```
      imageHandler.post(new Runnable() {
        @Override
        public void run() {
          if (imageView != null) {
            Log.d(LOG_TAG, "Image
            received");
            imageView.setImageBitmap(imageBitmap);
          }
        }
      });
    }
  }
}
};
```

5. Now, we need to create a new `loadBitmapFromAsset (Asset asset)` method and add the code that will load and decode the bitmap asset information, as shown in the following code snippet:

```
public Bitmap loadBitmapFromAsset(Asset asset) {
  if (asset == null) {
    throw new IllegalArgumentException("Asset cannot
      be empty");
  }
  // Convert asset into a file descriptor and block
  // until it's ready
  InputStream assetInputStream =
    Wearable.DataApi.getFdForAsset(mGoogleApiClient,
    asset).await().getInputStream();

  if (assetInputStream == null) {
    Log.w("WearActivity", "Requested an unknown
    Asset.");
    return null;
  }
  // Decode the stream into a bitmap
  return BitmapFactory.decodeStream(assetInputStream);
}
```

In the preceding code snippets, we start creating additional variables for imageView, imageBitmap, and imageHandler. The imageView will be used to output the contents of our image and display this to the Android wearable device. The imageBitmap will be responsible for holding the downloaded and decoded image, once it has been received by imageHandler. Next, we get a reference to our imageView object that will be used to display the image received from the Android handheld device. Then, we make a call to our initGoogleApiClient method that is responsible for establishing the connection between our Android handheld and wearable devices. Once the GoogleApiClient object has established a successful connection, the onConnected method is called where we set up a listener service on DataApi that will capture any images our wearable class receives, and these will be handled by the onDataChangedListener() method.

In our next step, we create a new onDataChangedListener() object that inherits from the DataApi.DataListener class that includes a onDataChanged method that will be responsible for receiving any images that are sent from the Android handheld device and accepts a dataEvents variable that contains the contents of our image data object. Next, we use the DataEvent class to get a list of all data objects that are currently being sent to the Android wearable device.

Then we use the getType property to get the type of event, and perform a comparison to check if there has been a change since the last time. We use the getPath property of the getDataItem property of the DataEvent event object to check that we are processing the correct image path value.

Next, we use the DataMapItem class to extract the contents of our image object from the event object, since we have obtained the correct data item path. Then we use the Asset class and use the getAsset method to get the image object data, which we pass to our loadBitmapFromAsset method to decode the contents of the image into a bitmap object and return the contents back to the calling method. We use the await property on our getFdForAsset method to block our wearable UI until the task completes. The imageHandler method is called only when an image object has been received, and then it uses the setImageBitmap method of our imageView object so that this can be displayed within the watch area, before finally destroying the connection to our GoogleApiClient within the onDestroy method.

Finally, we can compile, build, and run our application. Simply press *CMD + F9* and choose your AVD or Android handheld device from the list of Android emulators. Once the mobile app has been launched, you should see the user interface displayed, as shown in the following screenshot:

 You must ensure that you install the app on both your Android handheld device and the Android wearable device to ensure that everything works as expected.

The preceding screenshot shows the workflow between the Android handheld device and the Android wearable device when an image has been sent after the user taps the **Send an image to Android Wear** button.

Summary

In this chapter, we learned about three new APIs, `DataAPI`, `NodeAPI`, and `MessageAPIs`, and how we can use them and their associated methods to transmit information between the handheld mobile and the wearable.

 If, for whatever reason, the connected wearable node gets disconnected from the paired handheld device, the `DataApi` class is smart enough to try sending again automatically once the connection is reestablished.

We learned about `GoogleApiClient` that is responsible for establishing and handling the connection between the Android handheld and the wearable device. Then we moved on to learn about the `NodeApi` class, and how we can use this to obtain a list of all nodes that are currently connected to the wearable device. Next, we learned about `MessageApi` and how we can use the `sendMessage()` method to send a message to the Android wearable.

Then we got acquainted with the `DataApi` class, and how we can use the `DataListener()` method to check for any image data being sent from the Android handheld to the wearable.

In the next chapter, we will learn how we can build effective and interactive content for the Android TV platform. We will get acquainted with the Android Leanback support library to see how we can design and customize our own user interfaces.

5
Working with Google Glass

This chapter will provide you with the background and understanding of how you can effectively build applications that communicate directly with Google Glass and chances are, either you are most likely a Google Glass owner or you simply don't own one currently, but are intrigued by it. Google Glass is basically a headset that has a visual display on one side, making it more like a monocle, except that it has an optics pod display. The main input controls for Glass are voice, gestures, and touch, so when you begin building applications for Google Glass, you can use one or more of these forms of input.

Another reason is that the display is much smaller than handheld devices; you have less space to work with to show information, as a result of which user interfaces tend to be plain and simple. Google Glass comes with several built-in features, some of which allow users to take a picture by issuing a voice command or pressing the onboard shutter button, or winking using your right eye. You can also record videos by holding the shutter button for three seconds and then using the touchpad to extend the video timeframe to be more than 10 seconds. You also have the ability to activate speech recognition and voice dictation, as well as performing Google searches using your voice.

In this chapter, we will also learn how we can incorporate and make use of Google Glass voice capabilities to respond to and receive voice input from the wearable. At the end of the chapter, we will look at how we can incorporate and access the Glass camera to capture and save an image, and we will cover the Google Glassware principle design guidelines.

The following topics are covered in this chapter:

- Installing the Glass Development Kit preview
- Installing the Google USB drivers for Windows
- Creating and building a Google Glass application

- Receiving voice input with Google Glass
- Accessing camera with Google Glass
- Incorporating the Google Maps API with Google Glass
- Google Glassware principle design guidelines

Installing the Glass Development Kit preview

Before we can start developing applications for the Google Glass platform, we will need to look at how to install the Google Glass Development Kit SDK:

1. Launch **Android SDK Manager** using **SDK Manager** in Android Studio.

2. Click on the **SDK Platform** option and select the **Glass Development Kit Preview** packages that are located under the **Android 4.4.2 (API 19)** section to install them:

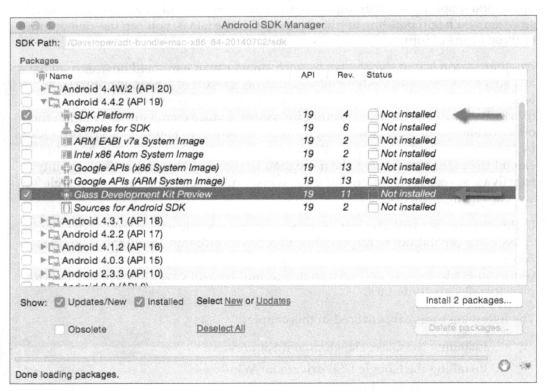

The **Google Glass Development Kit** (GDK) is currently in preview and is only available within **Android 4.4.2 (API 19) SDK**.

3. Next, accept the license information for **Android SDK License** and **Google Gdk License** by selecting the **Accept License** option.

4. Then, click on the **Install** button to begin installing the packages, as shown in the following screenshot:

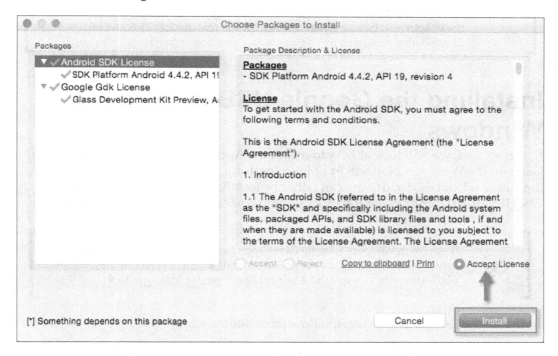

You will notice that the Glass Development Kit is still in developer preview and as such, at the time of writing, there is currently no emulator provided to test your apps, so you will need to have a Google Glassware device to deploy and test your apps.

If you do not own a Google Glass, but hope to learn how to develop for Google Glass, you can use an Android phone or tablet to test your Google Glass applications. Most of the user interfaces you create for Google Glass can be displayed normally on an Android phone or tablet, except that the voice trigger action provided by Google Glass will not work for phone and tablet.

Alternatively, there is a Google Glassware emulator by Gerwin Sturm called the Mirror API emulator. This works as a server-side API that means it doesn't run on Glass itself, but on your server and it's your server that interacts with Glass. This can act as a good idea of what the output will be like when it is run on the real device. This can be downloaded from `https://github.com/Scarygami/mirror-api`.

Installing the Google USB drivers for Windows

In this section, we will look at how to install the Android device drivers for Google Glass on the Windows platform in order to perform debugging with your Google devices. Android device drivers can either be installed from the Android SDK Google USB drivers or using the software that comes from the device manufacturer.

If you're developing Google Glass apps running on the Mac OS X or Linux platforms, you do not need to install a USB driver. On these platforms, you can use the Android File Transfer Manager tool.

To install the Google USB drivers, follow these simple steps:

1. Launch **Android SDK Manager** using **SDK Manager** in Android Studio.

2. Select and click on the **Google USB Driver** package, which is located under the **Extras** section, to install it:

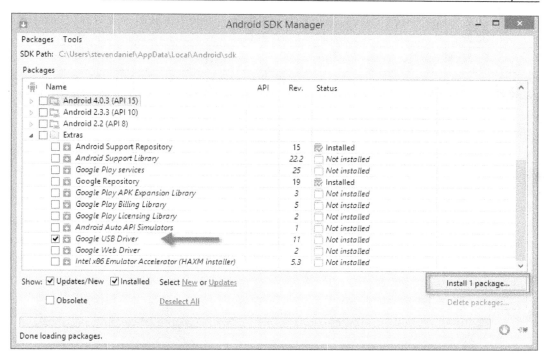

3. Next, click on the **install** button to begin installing **Google USB Driver**.

   Once you have installed the Google USB drivers for Google Glass, you will need to make a modification to the `android_winusb.inf` file, otherwise when you want to deploy your application's APK file on the device for debugging purposes, your device won't be listed, even though you have installed the drivers correctly.

4. Open the `android_winusb.inf` file that is located at `android_SDK_folder` under `\sdk\extras\google\usb_driver`, as shown in the following screenshot:

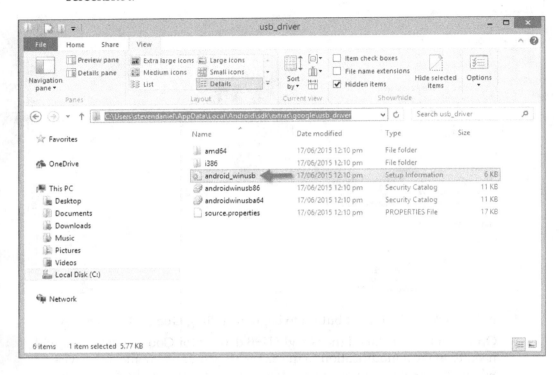

5. With the `android_winusb.inf` file displayed, we need to add the following `[Google.NTx86]` and `[Google.NTamd64]` sections as in the following snippet:

```
[Google.NTx86]

;GoogleGlass
%SingleAdbInterface% = USB_Install, USB\VID_18D1&PID_4E11&REV_0216
%CompositeAdbInterface% = USB_Install, USB\VID_18D1&PID_4E11&MI_01
%SingleAdbInterface% = USB_Install, USB\VID_18D1&PID_4E11&REV_0216
%CompositeAdbInterface% = USB_Install, USB\VID_18D1&PID_4E12&MI_01

[Google.NTamd64]
```

```
;GoogleGlass
%SingleAdbInterface% = USB_Install, USB\VID_18D1&PID_4E11&REV_0216
%CompositeAdbInterface% = USB_Install, USB\VID_18D1&PID_4E11&MI_01
%SingleAdbInterface% = USB_Install, USB\VID_18D1&PID_4E11&REV_0216
%CompositeAdbInterface% = USB_Install, USB\VID_18D1&PID_4E12&MI_01
```

6. Then launch the **Device Manager** application, right-click on your Google Glass device, and click to install the drivers. When you are prompted to browse for a location, select the `android_winusb.inf` parent folder, and follow the instructions presented.

 If you experience any issues with installing the Google USB driver, it is worthwhile to check out the documentation located at `http://developer.android.com/sdk/win-usb.html`.

Now that we have installed our Google Glass Development Kit and Android SDK, we can now start to build our Google Glass application for this chapter.

Creating and building a Google Glass application

In this section, we will look at how to create a native Google Glass wearable application that will enable us to communicate with our wearable device, so that we can create custom voice messages as well as access the Google Glass camera to take a photo and save the image to our wearable device.

Firstly, create a new project in Android Studio by following these simple steps:

1. Launch Android Studio, and then click on the **File | New Project** menu option.

2. Next, enter `HelloGoogleGlass` for the **Application name** field.

3. Then, provide the name for the **Company Domain** field.

4. Next, choose **Project location** where you would like to save your application code:

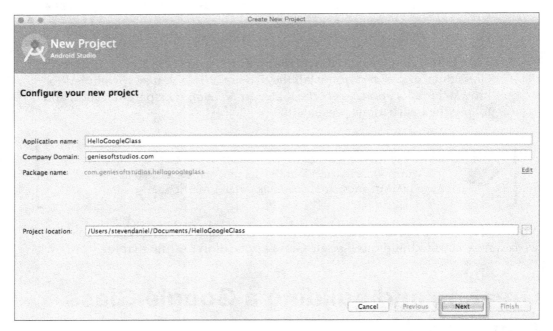

5. Finally, click on the **Next** button to proceed to the next step.

Next, we will need to specify the form factors for the Glass wearable device that our application will run on. On this screen, we will need to choose the minimum SDK version for the Glass wearable device.

6. Click on the **Glass** option and choose **Glass Development Kit Preview (Google Inc.)(API 19)** for the **Minimum SDK** option:

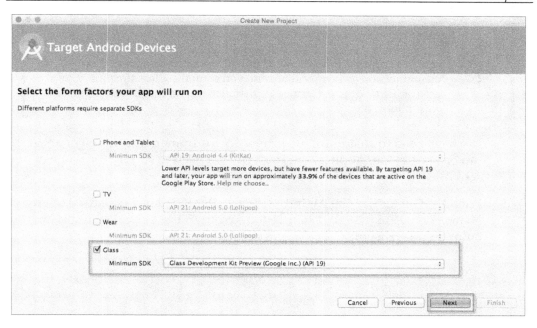

7. Click on the **Next** button to proceed to the next step.

8. From the **Add an activity to Glass** screen, choose the **Immersion Activity** option from the list of activities shown, and then click on the **Next** button to proceed to the next step:

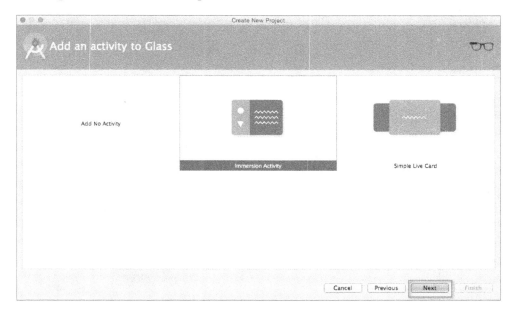

An immersion activity is the same as an Android activity. The name immersion is one that takes full control of the device. For example, when the user performs a swipe down gesture within Google Glass, the immersion activity will be removed from the Glass timeline. This type of behavior is performed in a similar way to those in the apps of your phone.

Next, we need to customize the properties for **Immersion Activity** so that it can be used by our application. Here we will need to specify the name of our activity as well as its title.

9. From the **Customize the Activity** screen, enter `MainActivity` for **Activity Name** and `Hello Glass Example` for **Activity Title** as shown in the following screenshot:

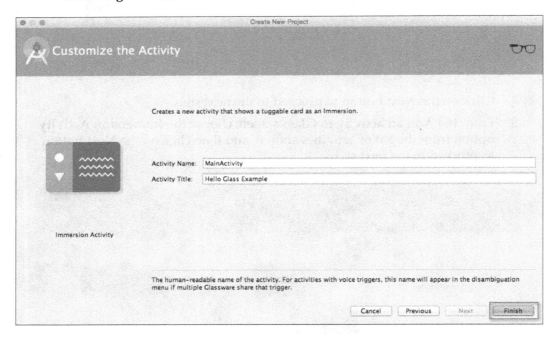

10. Next, click on the **Finish** button and the wizard will generate your project and after a few moments, the Android Studio window will appear with your project displayed in it.

Setting the theme for the Google Glass app

The next thing that we need to do is set up our application so that it doesn't take up too much of our Glass screen with the name of our immersion activity title bar, and certainly we don't want our screen to be gray with a black font. This is quite simple to fix.

From the **Project Navigator** window, open `styles.xml` that is located in the `res |` `values` folder, and modify this file by adding the following highlighted code:

```xml
<?xml version="1.0" encoding="utf-8"?>
<resources>
<style name="AppTheme" parent="android:Theme.DeviceDefault"></style>
</resources>
```

As you can see, we just need to switch the theme and let the Google Glass OS take care of it for us. All we did here is modified `android:Theme.Holo.Light` to `android:Theme.DeviceDefault` and this will automatically take care of any application layout and colors for us, using the default Glass theme.

Configuring the project to run on Google Glass

Before we can actually run the application on our Google Glassware, we must first ensure that we have enabled our device for testing and we need to configure our project to handle installation of our app on our Google Glass wearable:

1. On your Glassware device, go to **Settings** | **Device Info** | **Turn on debug**.

2. Next, plug your Google Glass into your computer's USB port. You will hear a dinging sound coming from your Google Glass to let you know that it has connected properly.

3. Now, we need to configure our app so that we can install this on our Google Glassware. Choose the app configuration from the drop-down menu, and choose the **Edit Configurations...** menu option as shown in the following screenshot:

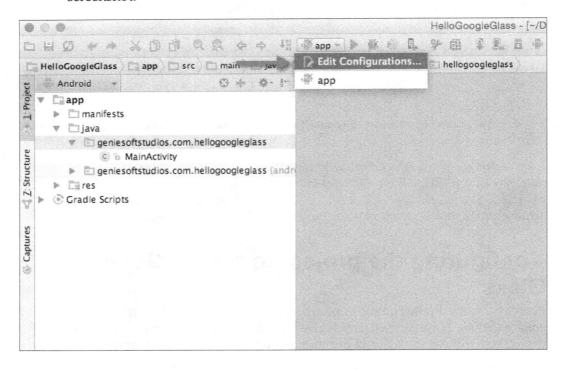

4. From the **Run/Debug Configurations** dialog, choose **Deploy default APK** under the **Package** section.

5. Under the **Activity** section, choose the **Launch** option and click on the **...** button to display the **Select Activity Class** dialog.

6. Next, choose the `MainActivity` class and then click on the **OK** button as follows:

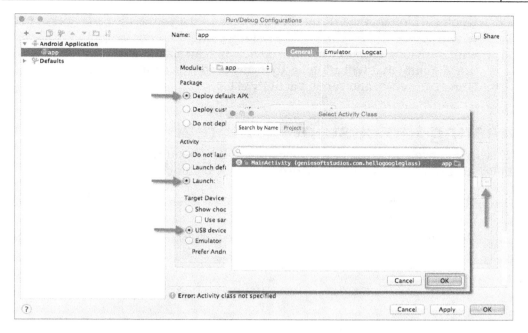

7. Under the **Target Device** section, choose the USB device and click on the **OK** button to dismiss the **Run/Debug Configurations** dialog as shown in this screenshot:

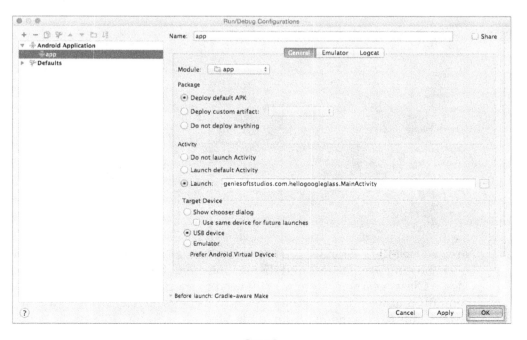

8. Upon clicking on the **OK** button, this will return you to the code editor window. Now that we have successfully configured our project to work with our Google Glass wearable device, we can start building additional functionality that will allow us to incorporate voice input so that it can launch our voice and camera activity classes.

Creating the custom menu resource file

In this section, we will proceed to create our custom menu resource file that will be used to store our custom voice keywords in order to launch the app handsfree:

1. From the **Project Navigator** window, expand the **app** section, then select and expand the **res** section.

2. Next, right-click and choose the **New** | **Android** resource file menu option, as shown in the following screenshot:

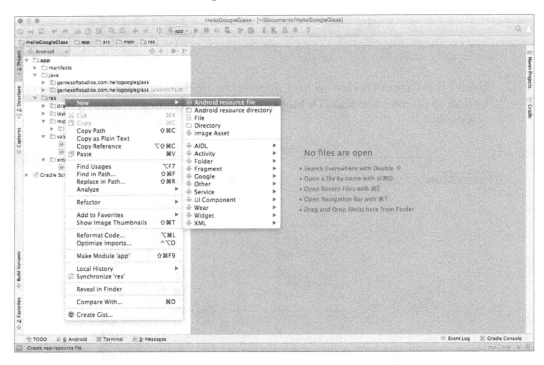

Next, we need to customize the properties for our resource file so that it can be used by our application, and display the menu items for use with our Google Glass wearable. Here we will need to specify the filename for our menu file and we need to specify **Resource type**, which tells Android that we need this to be a menu resource file.

3. Enter `activity_menu` for the **File name** field.

4. Next, choose the **Menu** item from the **Resource type** field drop-down menu, as shown in this screenshot:

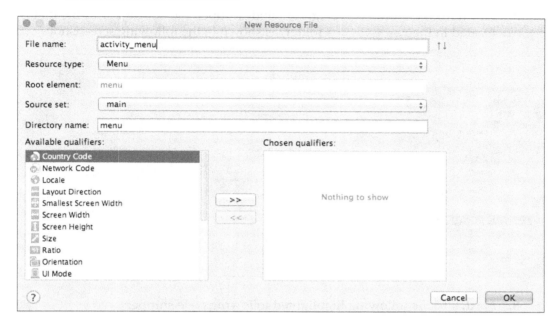

5. Next, click on the **OK** button to have the wizard generate the necessary files for you. Once finished, this will open the Android Studio code editor with your custom menu file displayed in it.

   Our next step is to construct the menu items that will appear within our Google Glass menu when launched.

6. From the **Project Navigator** window, open the `activity_menu.xml` file that is located in the `res | menu` folder, and add the following code:

```xml
<?xml version="1.0" encoding="utf-8"?>
<menu xmlns:android="http://schemas.android.com/apk/res/android">
  <item android:id="@+id/show_camera_item"
    android:title="@string/show_camera">
  </item>
  <item android:id="@+id/show_voice_item"
    android:title="@string/show_voice">
  </item>
  <item android:id="@+id/show_location_item"
    android:title="@string/show_location">
  </item>
</menu>
```

In our next step, we need to add the associated menu item's text values that will be displayed when our menu is displayed within Google Glass when launched.

7. Again from the **Project Navigator** window, open the `strings.xml` file that is located in the `res | values` folder, as shown in the following screenshot:

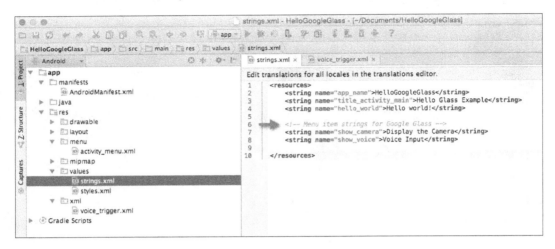

8. Next, add the following highlighted following code snippet:

```
<resources>
<string name="app_name">Hello Google Glass</string>
<string name="title_activity_main">Hello Glass
    Example</string>
<string name="hello_world">Hello world!</string>
<!-- Menu item strings for Google Glass -->
<string name="show_camera">Display the Camera</string>
<string name="show_location">Show my location</string>
<string name="show_voice">Voice Input</string>
</resources>
```

9. Now, we will need to change the currently declared `OK Glass, show me a demo` command that is currently located within our `voice_triggers.xml` file. Here we will need to include a contextual trigger that will be required to launch our app.

10. From the **Project Navigator** window, open the `voice_triggers.xml` file, which is located in the `res | values` folder, and modify this file by adding the following snippet:

```
<trigger keyword="@string/app_name" />
```

This changes the starting voice trigger to the same as your application's name; for example, the trigger is now `Hello Google Glass`.

 One thing to notice here is that we have replaced the command keyword with `keyword` so that we can use a contextual trigger now, that is, `OK Glass` followed by `Hello Google Glass`.

Configuring the AndroidManifests file

In our next section, we will need to make some additional changes to our `AndroidManifests` file, which will allow our app to support custom voice controls, as Google is pretty strict about which voice commands are allowed in approved Glass apps, and all new commands must be approved prior to your app being accepted on the Google Play Store:

1. From the **Project Navigator** window, select the `manifests` folder and then double-click on the `AndroidManifest.xml` file to open it.

2. Next, under the `manifest` section of the app, we need to include all permissions that will allow our app to support custom voice commands as well store images taken using the camera.

3. Enter the following highlighted code:

```xml
<?xml version="1.0" encoding="utf-8"?>
<manifest xmlns:android="http://schemas.android.com/apk/res/
android"
     package="geniesoftstudios.com.hellogoogleglass" >
<!-- Permissions to test new, unlisted voice commands-->
<uses-permission
  android:name="com.google.android.glass.permission.
  DEVELOPMENT"/>
<!-- Permissions to store our captured images -->
<uses-permission
android:name="android.permission.CAMERA"/>
<uses-permission
android:name="android.permission.READ_EXTERNAL_STORAGE"/>
<uses-permission
android:name="android.permission.WRITE_EXTERNAL_STORAGE"/>

<!-- For getting the current user location -->
<uses-permission android:name="android.permission.ACCESS_FINE_
LOCATION" />
<uses-permission android:name="android.permission.INTERNET" />
<application
```

```
            android:allowBackup="true"
            android:icon="@mipmap/ic_launcher"
            android:label="@string/app_name"
            android:theme="@style/AppTheme" >
<activity
            android:name=".MainActivity"
            android:icon="@drawable/ic_glass_logo"
            android:label="@string/title_activity_main" >
<intent-filter>
<action android:name="com.google.android.glass.action.VOICE_
TRIGGER" />
</intent-filter>
<meta-data
android:name="com.google.android.glass.VoiceTrigger"
            android:resource="@xml/voice_trigger" />
</activity>
<activity android:name=".CameraActivity"
            android:label="@string/app_name">
<intent-filter>
<action android:name="geniesoftstudios.com.hellogoogleglass.
CAMERA_ACTIVITY" />
<category android:name="android.intent.category.DEFAULT" />
</intent-filter>
</activity>
<activity android:name=".VoiceInputActivity"
            android:label="@string/app_name">
<intent-filter>
<action android:name="geniesoftstudios.com.hellogoogleglass.VOICE_
INPUT_ACTIVITY" />
<category android:name="android.intent.category.DEFAULT" />
</intent-filter>
</activity>
<activity android:name=".MapLocationActivity"
            android:label="@string/app_name">
<intent-filter>
<action android:name="geniesoftstudios.com.hellogoogleglass.MAP_
LOCATION_ACTIVITY" />
<category android:name="android.intent.category.DEFAULT" />
</intent-filter>
</activity>
</application>
</manifest>
```

In our next section, we will need to create a new layout resource file that will be used to display our camera preview when the camera activity is launched. We will be creating this file at this point and then begin adding the necessary code for our camera activity as we proceed through this chapter.

Creating the custom camera layout resource file

In this section, we will proceed to create our custom layout resource file that will be responsible for ensuring that our camera preview section displays correctly within our Google Glass wearable:

1. From the **Project Navigator** window, expand the **app** section and select and expand the **res | layout** section.

2. Next, right-click and choose the **New | Layout** resource file menu option as shown in the following screenshot:

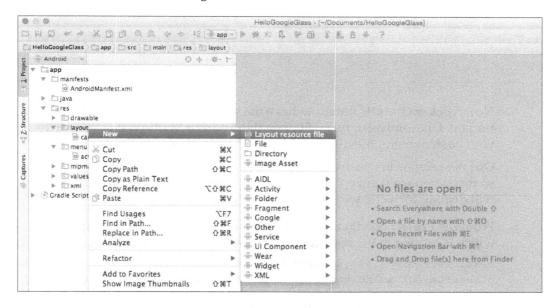

Now, we need to customize the properties for our resource file so that it can be used by our application. Here we will need to specify the filename for our resource file and we need to specify the root element name for our layout information.

3. Enter `camera_preview` for the **File name** field.

4. Next, enter `LinearLayout` for the **Root element** field as shown in this screenshot:

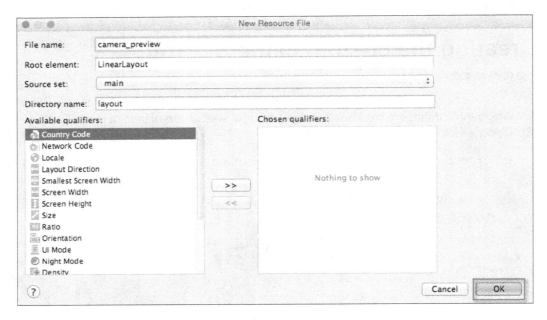

5. Now, click on the **OK** button to have the wizard generate the necessary files for you. Once finished, this will open the Android Studio code editor with your custom menu file displayed in it.

   Our next step is to construct the layout for our custom camera preview, so that when we launch the Glass camera functionality, our camera preview will display correctly within the Glass screen.

6. From the **Project Navigator** window, open the `camera_preview.xml` file, which is located in the **res | layout** section, and add the following code:

```xml
<?xml version="1.0" encoding="utf-8"?>
<LinearLayout xmlns:android="http://schemas.android.com/apk/res/
android"
    android:orientation="vertical"
    android:layout_width="fill_parent"
    android:layout_height="fill_parent">
<SurfaceView
        android:id="@+id/camerapreview"
        android:layout_width="fill_parent"
        android:layout_height="fill_parent"/>
</LinearLayout>
```

In our next section, we will look at how we can communicate with our Google Glass device by integrating custom voice actions and have our app launch the necessary activity based on the spoken word.

Incorporating a voice input within Google Glass

In this section, we will be looking at the steps involved in communicating with our Google Glassware wearable device to handle voice input. This process is quite simple and our next step is to write the code statements that will enable us to communicate with our Glassware wearable and display the user's spoken text:

1. From the **Project Navigator** window, expand the **app** section and select and expand the **java** section.

2. Next, right-click and choose the **New | Java Class** menu option:

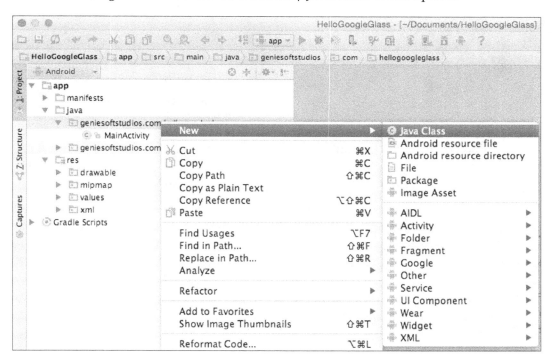

3. Then, enter `VoiceInputActivity` to be used as the name of our class and click on the **OK** button:

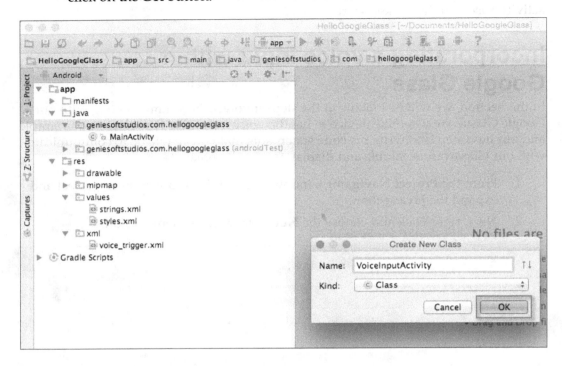

Upon clicking the **OK** button, the Android Studio code editor will open with our `VoiceInputActivity` class displayed. Our next step is to write the code that will be responsible for handling our voice, capturing our spoken voice, and displaying this within a card in the Glass user interface.

4. Open the `VoiceInputActivity.java` file that we just created.

5. Next, enter the following `import` statements::

```
import android.app.Activity;
import android.content.Intent;
import android.os.Bundle;
import android.speech.RecognizerIntent;
import android.view.View;
import com.google.android.glass.widget.CardBuilder;
import java.util.List;
import android.util.Log;
```

6. Now, we need to modify the VoiceInputActivity class by modifying the onCreate(Bundle savedInstanceState) method to include our receiveVoiceInput method that will start intent to ask the user for voice input when the class is instantiated, as shown in the following code:

```
public class VoiceInputActivity extends Activity {
        private final static int SPEECH_REQUEST = 0;
        @Override
        protected void onCreate(Bundle savedInstanceState) {
            super.onCreate(savedInstanceState);
            receiveVoiceInput();
```

7. Then, we need to create a new receiveVoiceInput method that will be called once the activity is instantiated and we listen for voice commands and decide what to do, as shown in this code snippet:

```
        public void receiveVoiceInput() {
// Start the intent to ask the user for voice input
Intent intent = new
Intent(RecognizerIntent.ACTION_RECOGNIZE_SPEECH);
        intent.putExtra(RecognizerIntent.EXTRA_PROMPT, "Speak
        now");
        startActivityForResult(intent, SPEECH_REQUEST);
}
```

8. Next, we need to create an onActivityResult method that will be called once the voice intent has returned and no further spoken text is detected. This is shown in the following code:

```
        @Override
        protected void onActivityResult(int requestCode, int
                            resultCode, Intent data) {
// When the voice input intent returns and is ok
if (requestCode == SPEECH_REQUEST && resultCode ==  RESULT_OK) {
// Get the text spoken
List<String> results = data.getStringArrayListExtra(
        RecognizerIntent.EXTRA_RESULTS);
        String spokenText = results.get(0);
        Log.d("VoiceInputActivity","text: "
        + spokenText);
// Add the text to the view so the user knows we
// retrieved it correctly
    CardBuilder card = new CardBuilder(this,
    CardBuilder.Layout.TEXT);
    card.setText(spokenText);
```

```
        View cardView = card.getView();
        setContentView(cardView);
    }
    super.onActivityResult(requestCode, resultCode, data);
  }
}
```

In the preceding code snippets, we started by adding our import statements that will be responsible for allowing our app to communicate with Google Glass. We added the `speech.RecognizerIntent` package to allow our class to receive speech input once the intent has been instantiated and then start listening for speech activity within our current class. The `onActivityResult` method is called when the speech recognizing intent completes and a result code of `RESULT_OK` or `RESULT_CANCEL` is returned.

In our next step, we check to see if we have received `resultCode` of `RESULT_OK` and then we get the spoken text by using its `EXTRA_RESULTS` from `RecognizerIntent`. Then we log the results to the debug window and update the Glass view to show the new data received.

Accessing camera through Google Glass

In this section, we will be looking at how we can access the Google Glassware camera, take a picture, and save it to the Glassware device:

1. From the **Project Navigator** window, expand the **app** section, then select and expand the **java** section.

2. Next, right-click and choose the **New | Java Class** menu option.

3. Then, enter `CameraActivity` to be used as the name for our class and click on the **OK** button.

   Our next step is to write the code that will be responsible for handling and capturing the image from the camera, and then saving this to the device's local storage.

4. Open the `CameraActivity.java` file that we just created.

5. Next, enter the `import` statements:
   ```
   import android.app.Activity;
   import android.content.Context;
   import android.graphics.Bitmap;
   import android.graphics.BitmapFactory;
   import android.graphics.Matrix;
   import android.graphics.PixelFormat;
   import android.hardware.Camera;
   ```

```
import android.media.AudioManager;
import com.google.android.glass.media.Sounds;
import android.os.Bundle;
import android.os.Environment;
import android.view.KeyEvent;
import android.view.SurfaceHolder;
import android.view.SurfaceView;
import android.util.Log;
import java.io.File;
import java.io.FileOutputStream;
import java.io.IOException;
import java.text.SimpleDateFormat;
import java.util.Date;
```

6. Here, we need to modify the `CameraActivity` class by modifying the `onCreate(Bundle savedInstanceState)` method that will initialize `Camera` when the class is instantiated, as shown in the following code snippet:

```
public class CameraActivity extends Activity {
    private SurfaceHolder _surfaceHolder;
    private Camera _camera;
    private boolean _previewOn;
    Context _context = this;
    @Override
    public void onCreate(Bundle savedInstanceState) {
super.onCreate(savedInstanceState);
        setContentView(R.layout.camera_preview);
// Set up the camera preview UI
getWindow().setFormat(PixelFormat.UNKNOWN);
        SurfaceView surfaceView = (SurfaceView)
    findViewById(R.id.camerapreview);
_surfaceHolder = surfaceView.getHolder();
_surfaceHolder.addCallback(new
SurfaceHolderCallback());
```

7. Now, we need to create a new `onKeyDown` method that will be called to start the intent when the user has tapped the touchpad to take a photo as shown in this code snippet:

```
    @Override
    public boolean onKeyDown(int keyCode, KeyEvent event){
        switch (keyCode) {
            // Google Glass TouchPad Tap
            case KeyEvent.KEYCODE_DPAD_CENTER:
            case KeyEvent.KEYCODE_ENTER: {
                Log.d("CameraActivity", "Tap.");
```

```
                    AudioManager audio =
                    (AudioManager)getSystemService(
                    Context.AUDIO_SERVICE);
                    audio.playSoundEffect(Sounds.SUCCESS);

                    // Take the picture
                    _camera.takePicture(null, null,
                    new SavePicture());
                    return true;
                }
                default: {
                    return super.onKeyDown(keyCode, event);
                }
            }
        }
    }
```

8. Next, we need to create a new `getFilename` method that will be called once the `onPictureTaken` callback method has completed, as shown in the following code:

```
// Create the image filename with the current timestamp
private String getFilename(boolean isThumbnail) {
    Log.d("CameraActivity", "Saving picture...");
    SimpleDateFormat sdf = new
      SimpleDateFormat("yyyyMMdd_HHmmss_SSS");

    // Build the image filename
    StringBuilder imageFilename = new StringBuilder();
    imageFilename.append(sdf.format(new Date()));
    if (isThumbnail) imageFilename.append("_tn");
    imageFilename.append(".jpg");

    // Return the full path to the image
    return
    Environment.getExternalStoragePublicDirectory
    (Environment.DIRECTORY_DCIM) +
      File.separator + "Camera" +
    File.separator + imageFilename;
}
```

9. Here, we need to create a new `savePicture` method that will be called once the `onPictureTaken` callback method has returned, as follows:

```
// Write the image to local storage
public void savePicture(Bitmap image, String filename) throws
IOException {
```

```
FileOutputStream fos = null;
try {
  fos = new FileOutputStream(filename);
  image.compress(Bitmap.CompressFormat.JPEG, 100, fos);
  Log.d("CameraActivity", "Picture saved.");
}
catch (IOException e) {
  e.printStackTrace();
  throw(e);
}
finally {
  fos.close();
}
}
```

10. Now, we need to create a new `surfaceChanged` method that will be called once the `SurfaceHolder` callback method has returned as shown in this code snippet:

```
// Handling of the camera preview
private class SurfaceHolderCallback implements SurfaceHolder.
Callback {
  @Override
  public void surfaceChanged(SurfaceHolder holder, int
    format, int width, int height) {
    if (null != _camera) {
      try {
        Camera.Parameters params =
          _camera.getParameters();
        params.setPreviewFpsRange(5000, 5000);
        _camera.setParameters(params);
        // Start the preview
        _camera.setPreviewDisplay(_surfaceHolder);
        _camera.startPreview();
        _previewOn = true;
      }
      catch (IOException e) {
        e.printStackTrace();
      }
    }
  }
  @Override
  public void surfaceCreated(SurfaceHolder holder) {
    _camera = Camera.open();
  }
```

```
      @Override
      public void surfaceDestroyed(SurfaceHolder holder) {
        if (_previewOn) {
          // Stop the preview and release the camera
          _camera.stopPreview();
          _camera.release();
        }
      }
    }
```

11. Next, we need to create a new method to handle the onPictureTaken callback method, as shown in the following code:

```
// Callback that is called when the picture is taken
class SavePicture implements Camera.PictureCallback {
  @Override
  public void onPictureTaken(byte[] bytes, Camera camera)
  {
    Log.d("CameraActivity", "Picture taken.");
    Bitmap image =
      BitmapFactory.decodeByteArray(bytes,
      0, bytes.length);
    try {
      // Save the image
      String imageFilename = getFilename(false);
      savePicture(image, imageFilename);
    }
    catch (IOException e) {
      e.printStackTrace();
    }
  }
}
```

In the preceding code snippets, we started by adding the import statements that will be responsible for allowing our app to communicate with Google Glass. Then we proceeded to create our private variables for SurfaceHolder and Camera as these will be responsible for displaying the camera preview within the Google Glass' screen window.

Then we call the setContentView method for our activity to use our custom camera_preview layout file once the activity is created, and then start setting up the callback methods for our camera preview UI. The onKeyDown method is called once the user has successfully tapped on TouchPad within the Google Glass control options, and then a sound effect notifying success is played through Glass to notify the user that the camera successfully took the picture.

The `onPictureTaken` callback method is called and this sends the picture as a byte array through which we use the `BitmapFactory.decodeByteArray` method to convert the bytes into a bitmap image. It is worth mentioning that in terms of your code, you shouldn't attempt to access the bytes of captured data, since this may not be immediately available.

In our next step, we call the `getFilename` method to build the filename of the image that will be saved to external storage. This filename will contain the current date and time appended to the end of the filename. Once we have the constructed filename, we proceed to call the `savePicture` method and to this, we pass in the image data and filename, which then writes a compressed version of the bitmap to the specified bitmap stream and compresses for maximum quality.

Incorporating the Google Maps API with Google Glass

In this section, we will be looking at how we can use the Google Maps API to obtain users' current location and address details, and plot their position on the map as a placeholder marker:

1. From the **Project Navigator** window, expand the **app** section, select and expand the **java** section.

2. Next, right-click and choose the **New | Java Class** menu option.

3. Then, enter `MapLocationActivity` to be used as the name of our class and click on the **OK** button.

   Our next step is to write the code that will be responsible for getting the current user's location and address information, and then use the Google Maps API to pass the derived latitude and longitude coordinates to create a map image in memory, which we will display to the Google Glass content view.

4. Open the `MapLocationActivity.java` file that we have just created.

5. Next, enter the `import` statements:

```
import java.io.IOException;
import java.io.InputStream;
import java.net.HttpURLConnection;
import java.net.URL;
import java.util.List;
import android.graphics.Bitmap;
import android.graphics.BitmapFactory;
import android.location.Address;
```

```
import android.location.Criteria;
import android.location.Geocoder;
import android.location.Location;
import android.location.LocationListener;
import android.location.LocationManager;
import android.os.AsyncTask;
import android.os.Bundle;
import android.app.Activity;
import android.content.Context;
import android.view.View;
import android.view.WindowManager;
import android.util.Log;
import com.google.android.glass.widget.CardBuilder;
```

6. Now, we need to modify the `MapLocationActivity` class by modifying the `onCreate(Bundle savedInstanceState)` method to start retrieving our current location when the class is instantiated, as shown in this code snippet:

```
public class MapLocationActivity extends Activity implements
LocationListener{
    public LocationManager mLocationManager;
    private CardBuilder card;

    @Override
    protected void onCreate(Bundle savedInstanceState) {
      super.onCreate(savedInstanceState);

      // Add the text to the view so the user knows
      // we retrieved it correctly.
      card = new CardBuilder(this, CardBuilder.Layout.TEXT);
      card.setText("Getting your location...");
      View cardView = card.getView();
      setContentView(cardView);

      // Request a static location from the location manager
      mLocationManager = (LocationManager)
        getSystemService(Context.LOCATION_SERVICE);

      // Set up a criteria object to get the location data,
      // using the GPS provider on the handheld device.
      Criteria criteria = new Criteria();
      criteria.setAccuracy(Criteria.ACCURACY_FINE);
      criteria.setAltitudeRequired(true);
```

```
List<String> providers =
  mLocationManager.getProviders(criteria, true);

// Asks the provider to send a location update
// every 10 seconds.
for (String provider : providers) {
  mLocationManager.requestLocationUpdates(provider,
10000, 10, this);
  }
  getWindow().addFlags(WindowManager.LayoutParams.
  FLAG_KEEP_SCREEN_ON);
  }
}
```

7. Here, we need to create an `onLocationChanged` method that will be called each time `LocationManager` determines that the user's location has changed, as shown in the following code:

```
// Method to show the current Latitude and Longitude
// to the user.
@Override
public void onLocationChanged(Location location) {
  // Get the Latitude and Longitude information
  String mLatitude =
    String.valueOf(location.getLatitude());
  String mLongitude =
    String.valueOf(location.getLongitude());
  // Attempt to get address information from the
  // static location object.
  Geocoder geocoder = new Geocoder(this);
  try {
    List<Address> addresses =
      geocoder.getFromLocation(location.getLatitude(),
      location.getLongitude(), 1);
    // Check to see if we have successfully
    // returned the address information.
    if (addresses.size() > 0) {
      Address mAddress = addresses.get(0);
      String mAddressInfo = "";
      for(int i = 0;
        i < mAddress.getMaxAddressLineIndex(); i++) {
        mAddressInfo += mAddress.getAddressLine(i)
        + " ";
      }
      // Display the address information within our Card
      card.setFootnote(mAddressInfo);
```

```
    }
  } catch (IOException e) {
    Log.e("LocationActivity", "Geocoder error", e);
  }

  // Then, call our google maps URL to return the map
  // for the latitude and longitude coordinates
  new LocationMapImageTask().execute(
  "http://maps.googleapis.com/maps/api/staticmap?"
    + "zoom=10&size=640x360&markers=color:green|" +
  mLatitude + "," + mLongitude);

  View cardView = card.getView();
  setContentView(cardView);
  }
```

8. Next, we need to create a new doInBackground method that will be called once the LocationMapImageTask callback method has returned from downloading the image asynchronously, as shown in this code:

```
// Private class to handle loading the Map and returning
// back an Bitmap image object.
private class LocationMapImageTask extends
  AsyncTask<String, Void, Bitmap> {

  @Override
  protected Bitmap doInBackground(String... stringURL) {
    Bitmap bmp = null;
    try {
      URL url = new URL(stringURL[0]);
      HttpURLConnection conn = (HttpURLConnection)
        url.openConnection();
      conn.setDoInput(true);
      conn.connect();
      InputStream inputStream = conn.getInputStream();
      BitmapFactory.Options options =
        new BitmapFactory.Options();
      bmp = BitmapFactory.decodeStream(inputStream,
        null, options);
    }
    catch (Exception e) {
      Log.e("LocationActivity",
        "LocationMapImageTask", e);
    }
    // Return the map as a bitmap image
    return bmp;
```

```
  }
  // After we have successfully executed our
  // doInBackground method we need to display our map image
  // to the card.
  @Override
  protected void onPostExecute (Bitmap result) {
    // Add the map image to our Google Glass card
    card.addImage(result);
    View cardView = card.getView();
    setContentView(cardView);
    super.onPostExecute(result);
  }
}
```

9. Now, as shown in the following code snippet, we need to create a new method to handle the LocationListener callbacks that will be used to determine when the location provider information has been enabled, disabled, or when the status changes:

```
@Override
public void onProviderDisabled(String arg0) {
  // Called when the provider is disabled by the user.
}
@Override
public void onProviderEnabled(String arg0) {
  // Called when the provider is enabled by the user.
}
@Override
public void onStatusChanged(String
  arg0,int arg1,Bundle arg2){
  // Called when the provider status changes.
}
```

In the preceding code snippets, we started by adding our import statements that will be responsible for allowing our app to communicate with Google Glass and obtain the user's current location information. Then we proceed to create our mLocationManager object that will be used to hold the location data and make a call to get a list of location providers. Then we call the requestLocationUpdates method to ask the provider to send a location update every ten seconds. We specify a value of ten meters for the third parameter that tells LocationListener to update only when the location has changed and also when the number of milliseconds specified by our second parameter of 10000 has passed.

For more information on the requestLocationUpdates property, as well as to know about choosing appropriate values for the second parameter that, which can help in conserving battery life, check out the documentation at http://developer.android.com/reference/android/location/LocationManager.html.

We specify a parameter value of ACCURACY_FINE for our Criteria object, which allows the app to access precise location information from location sources such as GPS, cell towers and Wi-Fi. Next, we create an onLocationChanged method, which will be called each time LocationManager determines that the user's location has changed. We extract the Latitude and Longitude values from our location object that is passed into this method by LocationListener and then use the Geocoder class, and the getFromLocation property that, returns an array of addresses based on the provided latitude and longitude values and display this information within the setFootnote property of our CardBuilder card object.

In our next step, we create an AsyncTask class and pass in the Google Maps URL along with the Latitude and Longitude values to the execute() method, which returns back a static map image from the doInBackground method that runs in the background and not on the main thread. Next, we use the BitmapFactory.decodeStream method, which will convert the bytes returned from the stream into a bitmap image. Once the doInBackground method has completed, the onPostExecute method runs on the main thread and is passed to the bitmap image in the result parameter, which we then add to our card using the addImage property.

Finally, we create our LocationListener methods such as onProviderDisabled, onProviderEnabled, and onStatusChanged. These methods are called when the provider service has been disabled by the user, enabled by the user, or when the status of the provider changes.

For more information on the LocationListener class, refer to the document at http://developer.android.com/reference/android/location/LocationListener.html.

Modifying the Google Glass main activity UI

In our next section, we will need to modify our MainActivity.java file so that based on the custom voice command issued, it will be able to launch the required custom activity. We will need to create a menu so that our menu items will be displayed within the **OK Glass** menu:

1. From the **Project Navigator** window, expand the **app** section, select and expand the **java** section.

2. Next, double-click to open the `MainActivity.java` file, and add the following highlighted code:

```
import com.google.android.glass.media.Sounds;
import com.google.android.glass.view.WindowUtils;
import com.google.android.glass.widget.CardBuilder;
import com.google.android.glass.widget.CardScrollAdapter;
import com.google.android.glass.widget.CardScrollView;
import android.app.Activity;
import android.content.Context;
import android.content.Intent;
import android.media.AudioManager;
import android.os.Bundle;
import android.view.Menu;
import android.view.MenuItem;
import android.view.View;
import android.view.ViewGroup;
import android.widget.AdapterView;
```

3. Now, we need to modify the `MainActivity` class by modifying the `onCreate(Bundle savedInstanceState)` method to include a reference to our `requestFeature` method. This will set up our app to use custom voice commands when the class is instantiated, as shown in the following code:

```
@Override
protected void onCreate(Bundle bundle) {
  super.onCreate(bundle);
  // Set up our app to use custom voice commands
  getWindow().requestFeature(WindowUtils.
  FEATURE_VOICE_COMMANDS);
```

4. Here, we need to create an `onCreatePanelMenu` method that will be called when `intent` has started, and will set up and display our custom menu options within our Google Glass menu that we specified within our `activity_menu.xml` file, as shown in this code snippet:

```
// Set up our menu options so that they will appear in the Google
Glass menu
@Override
public boolean onCreatePanelMenu(int featureId, Menu menu) {
  if (featureId == WindowUtils.FEATURE_VOICE_COMMANDS) {
    getMenuInflater().inflate(R.menu.activity_menu, menu);
    return true;
  }
  return super.onCreatePanelMenu(featureId, menu);Now
}
```

5. Next, we need to create an `onMenuItemSelected` method that will be called when `intent` has started, and will set up and display our custom menu options within our Google Glass menu that we specified in our `strings.xml` file, as shown in the following code:

```
// Method to call the relevant activity based on the voice command
public boolean onMenuItemSelected(int featureId, MenuItem item) {
    // Handle item selection
    if (featureId == WindowUtils.FEATURE_VOICE_COMMANDS) {
        if (item.getItemId() == R.id.show_camera_item) {
            // Open new activity to do camera preview
            Intent intent = new Intent(this, CameraActivity.
class);
            startActivity(intent);
            return true;
        } else if (item.getItemId() == R.id.show_voice_item) {
            // Open new activity to do voice input
            Intent intent = new Intent(this, VoiceInputActivity.
class);
            startActivity(intent);
            return true;
        } else if (item.getItemId() == R.id.show_location_item) {
            // Open new activity to show users location on the Map
            Intent intent = new Intent(this,
            MapLocationActivity.class);
            startActivity(intent);
            return true;
        }
    }

    return super.onMenuItemSelected(featureId, item);
  }
}
```

In the preceding code snippets, we started by adding our `import` statements that will be responsible for allowing the custom voice commands that we give to Google Glass to work correctly and as expected. Firstly, we need to tell our activity that we request permission for it to use custom voice commands that are handled by `getWindow().requestFeature(WindowUtils.FEATURE_VOICE_COMMANDS)`. Next, we create an `onCreatePanelMenu` method where we check whether we are listening for custom voice commands, and then we inflate our menu for our activity so that our voice command options appear within the **OK Glass** menu options.

After the user triggers a voice command, the `onMenuItemSelected` method is called and first it checks that `featureId` is for voice commands, which means that an item has been selected from the **OK Glass** menu. In our next step, when the voice commands are called, a check is performed to find out what the item is. Once this has been determined, the relevant `intent` is initialized and the `startActivity` method launches `intent`, displaying it on the Google Glass screen.

 When developing native apps on Glass, there are required patterns to use menus and stop the Glassware. Immersions requires that down swiping should exit the app and return the user to the timeline, whereas using live cards requires the presence of a dedicated `stop` command to terminate the app.

Launching the app within Google Glass

Next, we can finally begin to compile, build, and run our application. You must first ensure that you have connected your Google Glassware to your computer's USB port, then simply press *CMD + F9*, and your app will be installed on your Google Glassware wearable device.

When the app has been installed on your Android wearable device, it will be launched and from the **Hello Google Glass** screen you have to speak these words: `OK Glass...Hello Google Glass...Voice Input....` Then your app will recognize what you are asking and display the **Voice Activity** screen, which will contain your spoken words displayed within an individual card notification.

An introduction to GDK and the Google Mirror API

In this section, we will be discussing the Google Mirror API and how this is designed to give developers access to Glass development without any prerequisites, as this API is included as part of the core Glass OS, and does not rely on any third party development tools. Google provides developers with its cloud-based RESTful services that enable them to build Glassware apps by interacting with web service calls that are part of the Google API platform, and is fully hosted, managed, and maintained by Google.

The Mirror API works by allowing you to build web-based services that can interact with Google Glass by providing the functionality over a cloud-based API that does not require running code on a real Google Glass device.

When working with the Mirror API, you need to keep in mind the following things about this API:

- Executes an OAuth 2.0 request to obtain an authentication token
- Executes HTTP requests
- Provides the ability to post timeline items
- Receive notifications when your users interact with a timeline item, and receive location-based updates

When using the Mirror API, you need to keep in mind that every HTTP request sent from your Glass application needs to be authorized by providing a valid token with each request. Tokens are issued by the Google API, using the standard OAuth 2.0 protocol.

 For more information on using OAuth 2.0 to access Google APIs, refer to `https://developers.google.com/identity/protocols/OAuth2`.

If you work with the GDK, you should know that this is an extension of the Android SDK, which was initially designed to develop handheld device apps. This means, your Glassware can leverage the entire Android SDK right from its activities and services to obtaining the user's location using location-based services and the camera APIs.

The GDK provides the following functionality for Glass:

- Ability to launch activities in response to voice commands
- Ability to add cards to the timeline
- Access to widgets and views designed specifically for Glass that allows you to create layouts that are consistent with the rest of the platform

 For more information on the GDK, go to `https://developers.google.com/glass/develop/gdk/`.

The Mirror API playground

The Google Glass playground lets you experiment with how the content is displayed within Glass. You can use the playground to push content to your Glass wearable device, but if you don't own the real device, you can still see how this would look like on the Glass display:

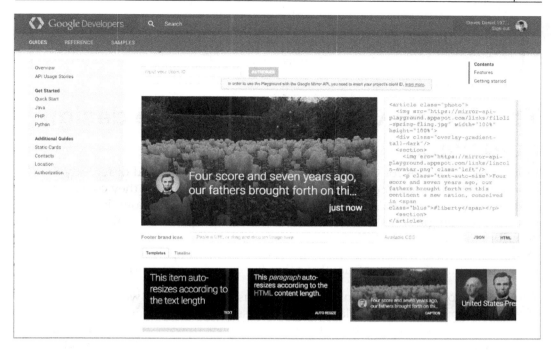

As you can see in the preceding screenshot, we can choose a sample template to use and see the associated HTML code that can be used to lay out our Glass content, so we can see how it would appear if we were running this on a real Glass wearable device.

The following code snippet shows the associated HTML code used in the preceding screenshot to display the content within our Glass display:

```html
<article class="photo">
  <img src="https://mirror-api-
    playground.appspot.com/links/filoli-spring-fling.jpg"
    width="100%" height="100%">
  <div class="overlay-gradient-tall-dark"/>
  <section>
    <img src="https://mirror-api
      -playground.appspot.com/links/lincoln-avatar.png"
      class="left"/>
    <p class="text-auto-size">Four score and seven years
      ago, our fathers brought forth on this continent a
      new nation, conceived in
      <span class="blue">#liberty</span></p>
  <section>
</article>
```

For more information on the Mirror API playground, please refer to the documentation at `https://developers.google.com/glass/tools-downloads/playground`.

The Google Glassware principle design guidelines

When designing apps for Google Glass, these need to be designed differently than how you would go about designing apps for phone or tablets, as they contain a different user experience. It is important to keep in mind and follow the Glassware principle design guidelines documentation that Google provides. This document describes the guidelines and principles that help you to design consistent user interfaces and user experiences for your Glass wearable apps, as well as ensure that your application runs efficiently within the Glassware platform.

You need to consider the screen sizes of your custom layouts, as well as the ease of use your app brings to the platform. The Glassware principle design guidelines document also explains how to ensure the consistency of your application while navigating from screen to screen, as well as the principles for designing good user interfaces are also covered. Some of the design pattern guidelines that your Glassware application needs to conform to are as follows:

- **Design for Glass**: When designing for Glass, users typically work with multiple devices that store and display information for a specific time period and Google Glass works best when information presented is simple, relevant, and current.

- **Don't get in the way**: Google Glass is designed to be there when you need it and it stays out of the way when you don't, but at the same time it must offer the same functionality that supplements the user's life without taking any of these features away.

- **Keep it relevant**: Information must be delivered at the right time, presented at the right place within the Google Glass user interface for each of your users, must contain information that is most relevant, and engage the users' attention and with the most satisfaction.

- **Avoid the unexpected**: Avoid sending content too frequently to your users, as this can have a bad experience for your customers. This is because Google Glass is so close to the users' senses that sending information too frequently and at unexpected times will frustrate your users. Always make it clear to users what is the intention of your Glassware and never pretend to be something you're not.

- **Build for people**: When designing your users' interfaces, make sure that you use imagery, colloquial voice interactions, and natural gestures. Ensure that you focus on a fire-and-forget usage model where users can start actions quickly and continue with what they're doing.

There is also information relating to the proper use and appearance of UI elements, such as background images and icons, as well as how to go about distributing your app to the Google Glassware platform.

 To obtain further information about these guidelines, it is worth checking out the Google Glassware principle guidelines documentation at https://developers.google.com/glass/design/principles.

Summary

In this chapter, we learned about Google Glass and how we can use this platform to build effective and interactive content by designing custom layouts to display content on the Google Glass screen. We learned how to install and set up the Google Glass Development Kit preview SDK, and got a brief overview about the Mirror API emulator since there is no emulator currently for Glass development. However, there is currently a Google Mirror API playground, which is a test environment to work with static timeline cards. Then we had an introduction to the two different activities that can be created for Glass, live cards, and immersions:

- Live cards are basically activities that are added to the Glass timeline and display information relating to high-frequency updates in real-time, which are constantly run in the background even when users are interacting with different cards. These activities allow users to multitask and access different kinds of information in real-time.

- Immersions, on the other hand, are fully customizable screens that run outside of the timeline experience. Using this type of activity allows you to design your own user interfaces and process user input.

Next, we learned how to customize the appearance and theme of custom menus as they appear within the **OK Glass** menu and how we can incorporate various voice commands within our app. We also learned about the CardScrollView class, which is a container that displays multiple cards side-by-side, whereas Card displays the content on screen. In addition, we learned how we can work with the Glass camera to take a photo and save it as an image on the Google Glass device.

In the next and final chapter, we will learn how to create and customize layouts and use layout themes for Android TV, as well as create the necessary activity fragments to display content.

6
Designing and Customizing Interfaces for Android TV

This chapter will provide you with the background and understanding of how you can effectively design and customize user interfaces for Android TV. When Google announced the Android TV platform during their Google I/O conference back in 2014, their vision was to create a highly interactive and connected television experience that would leverage and build upon the existing functionality found in the Android platform.

Google also provided consumers with the choice of either purchasing a smart TV with the platform built in, or alternatively adding Android TV to their existing television set by purchasing a stand-alone set-top box, such as the Nexus Q.

Android TV essentially brings apps and functionality that users already enjoy working with on their smaller Android devices, but with the added ability to download Android TV apps from the Google Play Store. It also provides users with a platform that supports Google Cast that will enable them to cast content from their smartphone or tablet onto their Android TV device to make the viewing experience much more exciting and usable from a living room couch.

This chapter includes the following topics:

- Creating and building an Android TV application
- Learning how to check for the presence of an Android TV interface
- Learning how to design and customize user interfaces
- Learning how to implement search functionality within Android TV
- Introducing you to the Android TV user interface guidelines

Creating and building an Android TV application

In this section, we will take a look at how to create an Android TV application using the default Android TV template that Android Studio provides us with. As we progress through the chapter, we will learn how to customize the Android TV user interface by creating our very own custom fragment classes to display header and row content information within the Android TV user interface.

 In Android, a fragment is a class that represents a behavior or portion of the user interface within an activity. Fragments were introduced to help produce the user interface, so that it can adapt to the various device orientations as well as function seamlessly across phones and tablets. You can even use multiple fragments within the same activity and rearrange them when the user rotates their device.

Firstly, create a new project in Android Studio by following these simple steps:

1. Launch Android Studio, and then click on the **File | New Project** menu option.

2. Next, enter `AndroidTVInterfaces` for the **Application name** field.

3. Then, provide the name for the **Company Domain** field.

4. Now, choose the project location where you would like to save your application code:

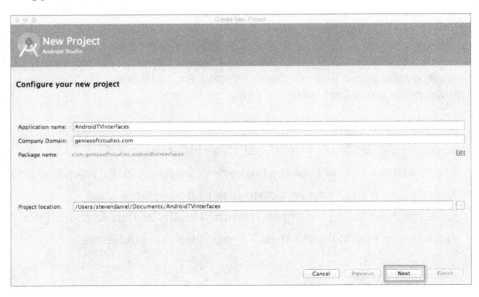

5. Click on the **Next** button to proceed to the next step.

   Next, we will need to specify the target form factors of our Android devices that our application will run on. On this screen, we will need to choose the minimum SDK version for our TV.

6. Click on the **TV** option and choose the **API 21: Android 5.0 (Lollipop)** option for **Minimum SDK**:

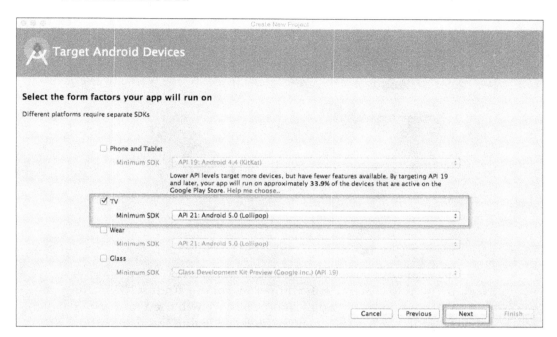

7. Click on the **Next** button to proceed to the next step.

   In our next step, we will need to add **Android TV Activity** to our application project for the TV section of our app.

8. From the **Add an activity to TV** screen, choose the **Android TV Activity** option from the list of activities shown and click on the **Next** button to proceed to the next step:

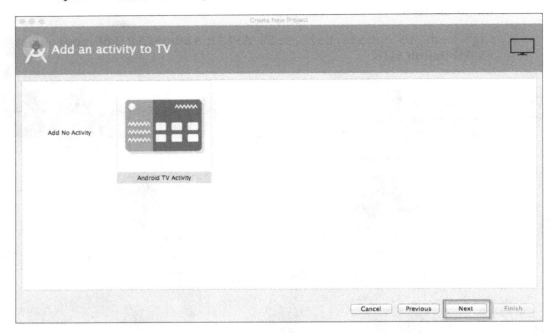

Next, we need to customize the properties for our Android TV activity so that it can be used by our application. Here we will need to specify the name for our activity, layout information, title, and layout fragment files.

9. From the **Customize the Activity** screen, enter MainActivity for **Activity Name** shown and click on the **Next** button to proceed to the next step in the wizard:

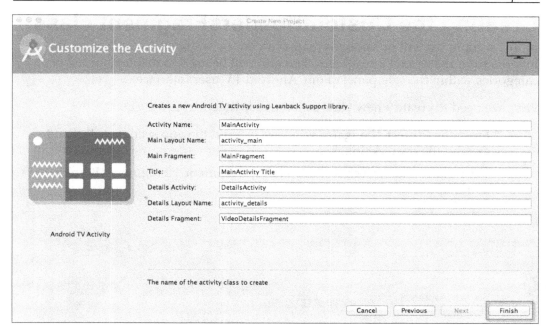

10. Finally, click on the **Finish** button and the wizard will generate your project and after a few moments, the Android Studio window will appear with your project displayed in it.

In our next section, we will take a look at how we can create activity and fragment classes to customize the look and feel of an Android TV user interface.

Customizing the Android TV user interface

In this section, we will begin by creating a custom header and custom row fragment for our Android TV user interface. The previously generated code that handles the displaying of header and row information currently exists within the MainFragment class.

In the sections that follow, we will be taking a look at how we can separate this information into two individual classes, which will make the code easier to maintain.

Creating the CustomHeadersFragment class

In this section, we will proceed to create our CustomHeadersFragment class that inherits from the HeadersFragment class and will be used to display our list of categories within the side panel of our Android TV user interface.

First, we need to create a new class called CustomHeadersFragment:

1. From the **Project Navigator** window, expand the **app** section, select and expand the **java** section.

2. Next, right-click and choose the **New | Fragment | Fragment (Blank)** menu option:

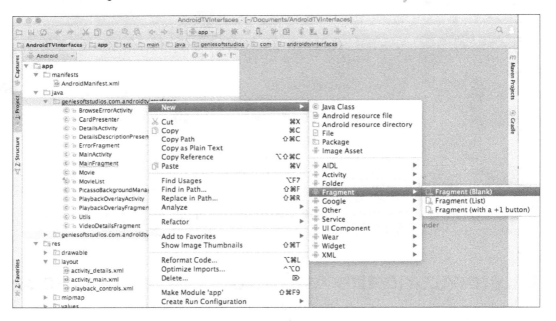

3. Then, enter CustomHeadersFragment to be used as the name for **Fragment Name**.

4. Next, ensure that you have not selected **Create layout XML?**.

5. Now, ensure that the **Include fragment factory methods?** and **Include interface callbacks?** options have not been selected.

6. Then click on the **Finish** button:

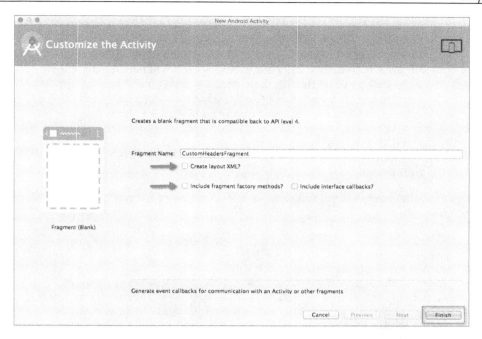

Upon clicking the **Finish** button, the Android Studio code editor will open as shown in the following screenshot:

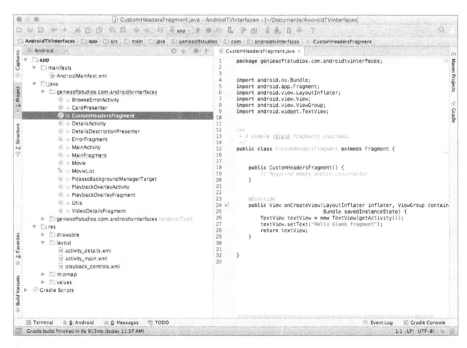

Over time, Google may decide to make changes to the Android Support Libraries for a Leanback support and you may notice that certain libraries and method calls don't exist, so you will need to change the code slightly to work with these libraries. Please refer to the Support Library document at `https://developer.android.com/tools/support-library/index.html#revisions` to help you.

Our next step is to write the code that will populate the category details within our Android TV side panel:

1. Open the `CustomHeadersFragment.java` file as shown in the preceding screenshot.

2. Next, enter the following `import` statements:

```
import android.app.Fragment;
import android.os.Bundle;
import android.support.v17.leanback.app.HeadersFragment;
import android.support.v17.leanback.widget.ArrayObjectAdapter;
import android.support.v17.leanback.widget.HeaderItem;
import android.support.v17.leanback.widget.ListRow;
import android.util.Log;
import android.view.View;
import android.widget.AdapterView;
import java.util.LinkedHashMap;
```

3. Now, modify the `CustomHeadersFragment` class by handling the `onActivityCreated` callback method that will be called when the fragment is instantiated, as shown in this code snippet:

```
public class CustomHeadersFragment extends HeadersFragment {
  private static final String TAG = "CustomHeadersFragment";
  private ArrayObjectAdapter mAdapter;
  @Override
  public void onActivityCreated(Bundle savedInstanceState)
    {
    Log.i(TAG, "onCreate");
    super.onActivityCreated(savedInstanceState);

    setHeaderAdapter();
    setCustomPadding();
  }
```

4. Then, we need to create a `setHeaderAdapter` method that will be called to display the category names within our side panel as shown in the following code:

```
private void setHeaderAdapter() {
  mAdapter = new ArrayObjectAdapter();
```

```
LinkedHashMap<Integer, CustomRowsFragment>
  fragments = ((MainActivity) getActivity()).getFragments();
int id = 0;
for (int i = 0; i < fragments.size(); i++) {
  HeaderItem header = new HeaderItem(id, "Category " + i);
  ArrayObjectAdapter innerAdapter = new
  ArrayObjectAdapter();
  innerAdapter.add(fragments.get(i));
  mAdapter.add(id, new ListRow(header, innerAdapter));
  id++;
}
setAdapter(mAdapter);
}
```

5. Next, we need to create a `setCustomPadding` new method that will be called to adjust the display when the categories are displayed within the side panel as shown in this code snippet:

```
private void setCustomPadding() {
  getView().setPadding(0,
  Utils.convertDpToPixel(getActivity(), 128),
  Utils.convertDpToPixel(getActivity(), 48), 0);
}
```

6. Finally, we need to create an `OnItemSelectedListener` method that will be called when an item has been selected within the side panel window as shown in the following code:

```
private AdapterView.OnItemSelectedListener
getDefaultItemSelectedListener() {
  return new AdapterView.OnItemSelectedListener() {
    @Override
    public void onItemSelected(AdapterView<?>
      adapterView, View view, int i, long l) {
      Object obj = ((ListRow)
        adapterView.getItemAtPosition(i))
        .getAdapter().get(0);
      getFragmentManager().beginTransaction().
        replace(R.id.rows_container,
        (Fragment) obj).commit();
      ((MainActivity)
        getActivity()).updateCurrentRowsFragment
        ((CustomRowsFragment) obj);
    }
    @Override
```

```
            public void onNothingSelected(AdapterView<?> adapterView) {
                Log.d(TAG,"Nothing has been selected");
            }
        };
    }
```

In the preceding code snippets, we started by adding our import statements that will be responsible for allowing our application to communicate with Android TV. We incorporate the Leanback support library that provides us with prebuilt components for our TV interface. We then proceed to extend our CustomHeadersFragment class using the HeadersFragment class and then add the code for our onActivityCreated (Bundle savedInstanceState) method. This will be called when the activity is instantiated and sets up a setOnItemSelectedListener listener method that will be responsible for the category after it has been selected.

In our next step, we create a setHeaderAdapter method that will be called to populate the category items in the left side panel. This method creates an ArrayObjectAdapter class that contains a list of all of our header items and will be called each time a category has been chosen and calls the CustomRowsFragmentListRow element to retrieve the associated row information for the chosen category. In our next step, we create a setCustomPadding method that will be used to adjust the padding for our fragment view, as soon as it is created.

Creating the CustomRowsFragment class

In this section, we will proceed to create our custom rows fragment class that inherits from the RowsFragment class and will be used to display our row information for the selected category that has been clicked within the side panel of our Android TV user interface.

First, we need to create our CustomRowsFragment fragment like we did in the previous section:

1. From the **Project Navigator** window, expand the **app** section, select and expand the **java** section.

2. Next, right-click and choose the **New** | **Fragment** | **Fragment (Blank)** menu option and enter CustomRowsFragment to be used as the name for **Fragment Name**.

3. Now, ensure that you have not selected the **Create Layout XML?** option.

4. Then, ensure that the **Include fragment factory methods?** and **Include interface callbacks?** options have not been selected and then click on the **Finish** button to open the Android Studio code editor window.

Our next step is to write the code that will be responsible for populating our row information within the Android TV interface.

5. Open the `CustomRowsFragment.java` file that we just created.

6. Next, enter the following `import` statements:

```
import android.graphics.Color;
import android.os.Bundle;
import android.support.v17.leanback.app.RowsFragment;
import android.support.v17.leanback.widget.ArrayObjectAdapter;
import android.support.v17.leanback.widget.HeaderItem;
import android.support.v17.leanback.widget.ListRow;
import android.support.v17.leanback.widget.ListRowPresenter;
import android.util.Log;
import android.util.TypedValue;
import android.view.LayoutInflater;
import android.view.View;
import android.view.ViewGroup;
import java.util.Collections;
import java.util.List;
```

7. Then, modify the `onCreateView(LayoutInflater inflater, ViewGroupcontainer, Bundle savedInstanceState)` method that will be called when the fragment is created, as shown in the following code:

```
public class CustomRowsFragment extends RowsFragment {
    private final int NUM_ROWS = 5;
    private final int NUM_COLS = 15;
    private ArrayObjectAdapter rowsAdapter;
    private CardPresenter cardPresenter;
    private static final int HEADERS_FRAGMENT_SCALE_SIZE = 300;
    private static final String TAG = "CustomRowsFragment";
    @Override
    public View onCreateView(LayoutInflater inflater, ViewGroup
      container, Bundle savedInstanceState) {
        View v = super.onCreateView(inflater, container,
          savedInstanceState);
        int marginOffset = (int)
          TypedValue.applyDimension(TypedValue.COMPLEX_UNIT_DIP,
          HEADERS_FRAGMENT_SCALE_SIZE,
          getResources().getDisplayMetrics());
```

```
    ViewGroup.MarginLayoutParams params =
        (ViewGroup.MarginLayoutParams) v.getLayoutParams();
    params.rightMargin -= marginOffset;
    v.setLayoutParams(params);
    v.setBackgroundColor(Color.DKGRAY);
    return v;
}
```

8. Next, modify the `CustomRowsFragment` class by creating an `onActivityCreated (Bundle savedInstanceState)` method that will be called when the fragment is instantiated, as shown in the following code snippet:

```
@Override
public void onActivityCreated(Bundle savedInstanceState) {
  Log.i(TAG, "onCreate");
  super.onActivityCreated(savedInstanceState);
  loadRows();
  setCustomPadding();
}
```

9. Now, as shown in the following code, we need to create a `loadRows()` new method that will be called to display the associated row information for the chosen category within our side panel:

```
private void loadRows() {
  rowsAdapter = new ArrayObjectAdapter(new
    ListRowPresenter());
  cardPresenter = new CardPresenter();
  List<Movie> list = MovieList.setupMovies();
  int i;
  for (i = 0; i < NUM_ROWS; i++) {
    if (i != 0) Collections.shuffle(list);
    ArrayObjectAdapter listRowAdapter = new
      ArrayObjectAdapter(cardPresenter);
    for (int j = 0; j < NUM_COLS; j++) {
      listRowAdapter.add(list.get(j % 5));
    }
    HeaderItem header = new HeaderItem(i,
      MovieList.MOVIE_CATEGORY[i]););
    rowsAdapter.add(new ListRow(header, listRowAdapter));
  }
  setAdapter(rowsAdapter);
}
```

10. Then, we need to create a `setCustomPadding()` method that will be called to adjust the padding for each of the rows within our Android TV interface as follows:

```
private void setCustomPadding() {
  getView().setPadding(Utils.convertDpToPixel(getActivity(),
    -24), Utils.convertDpToPixel(getActivity(), 128),
    Utils.convertDpToPixel(getActivity(), 48), 0);
}
```

11. Finally, we need to create a new `refresh()` method that will be called to adjust the padding for each of the rows within our Android TV interface when the contents have changed as shown in the following code snippet:

```
public void refresh() {
  getView().setPadding(Utils.convertDpToPixel(getActivity(),
    -24), Utils.convertDpToPixel(getActivity(), 128),
    Utils.convertDpToPixel(getActivity(), 300), 0);
}
```

In the preceding code snippets, we started by adding our import statements that will be responsible for allowing our application to communicate with Android TV and, just as we did in our `CustomHeadersFragment` class, we incorporate the Leanback support library that provides us with prebuilt components for our TV interface.

We then proceed to extend our `CustomRowsFragment` class using the `RowsFragment` class and then add the code for our `onCreateView (LayoutInflater inflater, viewGroup container, Bundle savedInstanceState)` method that will be called when the view has been created. This is responsible for setting the layout information for our margins and background of our fragment. In our next step, we add the code for our `onActivityCreated (Bundle savedInstanceState)` method that will be called when the activity is instantiated and calls the `loadRows` method to populate our fragment with information for the corresponding chosen category, before calling the `setCustomPadding` method that will be used to adjust the padding for our fragment view as soon as it is created.

In our next step, we create a `loadRows` method that will be called each time the activity is created and populates our view fragment with the associated row information for the chosen category. This method sets up an `ArrayObjectAdapter` class that instantiates the `ListRowPresenter` object, and then calls the set `upMovies` method from our `MovieList` class model and assigns this to a list object.

Next, we iterate through each row and column, shuffle the contents of our list object to ensure that we get different information each time our category is selected, create a listRowAdapter object that inherits from the CardPresenter class, extract the movie header information, and add the item details to rowAdapter for the chosen category. In our next step, we create a setCustomPadding method and refresh that will be used to adjust the padding for our fragment view, as soon as it is created.

Creating the CustomFrameLayout class

In this section, we will proceed to create our custom frame layout class that inherits from the FrameLayout class and will be used to ensure that the information is presented correctly within the TV interface.

First, we need to create our CustomFrameLayout class like we did in the previous section:

1. From the **Project Navigator** window, expand the **app** section, select and expand the **java** section.

2. Next, right-click and choose the **New | Java Class** menu option:

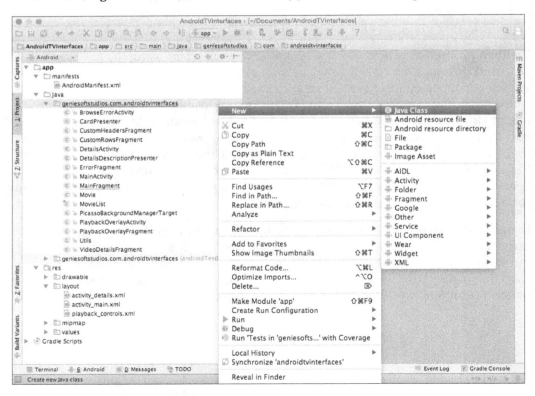

3. Then, enter `CustomFrameLayout` to be used as the name for our class and click on the **OK** button:

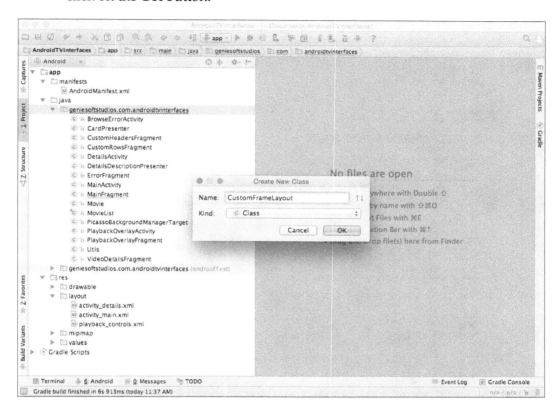

Upon clicking the **OK** button, the Android Studio code editor will open, as shown in the following screenshot:

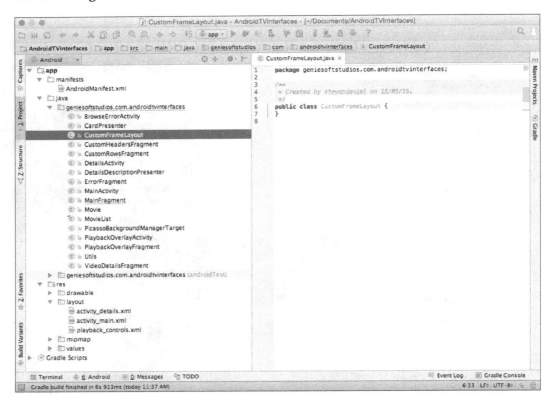

Our next step is to write the code that will be responsible for ensuring that the layout of our custom header and row information renders correctly when it is being displayed within the Android TV interface:

1. Open the `CustomFrameLayout.java` file that we just created.

2. Next, enter the following `import` statements:

```
import android.content.Context;
import android.util.AttributeSet;
import android.widget.FrameLayout;
```

3. Then, modify the `CustomFrameLayout` class as shown in the following code snippet:

```
public class CustomFrameLayout extends FrameLayout {
  public CustomFrameLayout(Context context) {
    this(context, null, 0);
  }
  public CustomFrameLayout(Context context,
    AttributeSet attrs) {
    this(context, attrs, 0);
  }
  public CustomFrameLayout(Context context, AttributeSet
    attrs, int defStyle) {
    super(context, attrs, defStyle);
  }
}
```

In the preceding code snippets, we started by adding our `import` statements that will be responsible for allowing our application to communicate with Android TV. We then proceed to extend our `CustomFrameLayout` class using the `FrameLayout` class, which is used to handle the positioning of all child items within the view. Next, we create our class constructor and add the code for our `CustomFrameLayout(Content context)` method that will be called when the class is instantiated and called. The additional overloaded methods are required when inheriting from the `FrameLayout` class, and are there for handling the setting of attributes, default layout styles, and so on.

Creating the SearchActivity class

In this section, we will proceed to create our custom `SearchActivity` class that will enable us to search for content using Google Play Services. This class will call a custom fragment class, which we will be creating later on.

First, we need to create a new blank `SearchActivity` class, which is basically an application component that will provide us with a screen so that the users can interact with it:

1. From the **Project Navigator** window, expand the **app** section, select and expand the **java** section.

2. Next, right-click and choose the **New | Activity | Blank Activity** menu option as shown in the following screenshot:

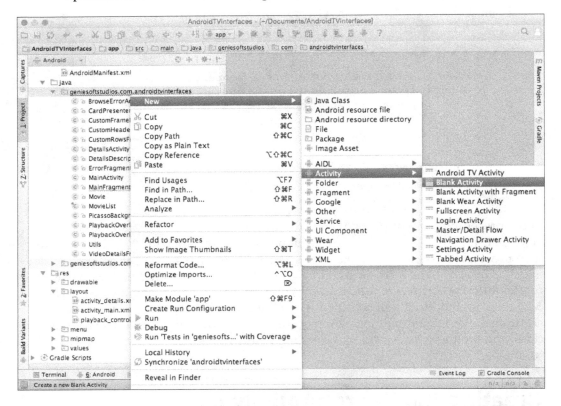

Next, we need to customize the properties for our blank activity so that it can be used by our application. Here we will need to specify the name for our activity, the layout information, and the title for our activity.

3. Enter `TVSearchActivity` for the **Activity Name** option and `activity_search` for **Layout Name**.

4. Next, enter `TVSearchActivity` for **Title** as shown in this screenshot:

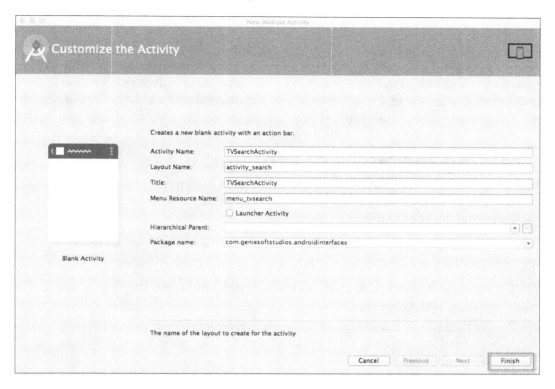

5. Then, click on the **Finish** button to have the wizard generate the necessary files for you. Once finished, this will open the Android Studio code editor with your code file displayed in it.

Our next step is to write the code that will be responsible for calling the layout information and our fragment file that will be used to handle our search functionality:

1. Open the `TVSearchActivity.java` file that we just created.

2. Next, enter in the following `import` statements:

    ```
    import android.app.Activity;
    import android.os.Bundle;
    ```

3. Now, modify the `TVSearchActivity` class as shown in the following code:

    ```
    public class TVSearchActivity extends Activity {
      @Override
      protected void onCreate(Bundle savedInstanceState) {
        super.onCreate(savedInstanceState);
    ```

```
    setContentView(R.layout.activity_search);
  }
}
```

In the preceding code snippets, we started by adding our `import` statements that will be responsible for allowing our application to communicate with an Android TV. We then proceeded to add the code for our `onCreate(Bundle savedInstanceState)` method that will be called when the class is instantiated and then display our search layout content within the current view.

Creating the SearchFragment class

In this section, we will proceed to create our custom search fragment class that will be inherited from the `SearchFragment` class. This class will be used to handle all of the search capabilities with our customized TV user interface.

First, we need to create our `TVSearchFragment` fragment class like we did in the previous sections:

1. From the **Project Navigator** window, expand the **app** section, then select and expand the **java** section.

2. Next, right-click and choose the **New | Fragment | Fragment (Blank)** menu option and enter `TVSearchFragment` to be used as the name for our fragment.

3. Next, ensure that you have not selected the **Create Layout XML?** option.

4. Now, ensure that the **Include fragment factory methods?** and **Include interface callbacks?** options have not been selected and then click on the **Finish** button to open the Android Studio code editor window.

Our next step is to write the code that will be responsible for handling all of the search capabilities within our customized Android TV interface:

1. Open the `TVSearchFragment.java` file that we just created.

2. Next, enter the `import` statements as shown in the following code:

```
import android.os.Bundle;
import android.os.Handler;
import android.support.v17.leanback.app.SearchFragment;
import android.support.v17.leanback.widget.ArrayObjectAdapter;
import android.support.v17.leanback.widget.HeaderItem;
import android.support.v17.leanback.widget.ListRow;
import android.support.v17.leanback.widget.ListRowPresenter;
import android.support.v17.leanback.widget.ObjectAdapter;
```

```
import android.support.v17.leanback.widget.OnItemClickedListener;
import android.support.v17.leanback.widget.Row;
import android.text.TextUtils;
import android.util.Log;
```

3. Now, modify the `TVSearchFragment` class as shown in this code snippet:

```
public class TVSearchFragment extends SearchFragment implements
SearchFragment.SearchResultProvider {
    private static final String TAG = "TVSearchFragment";
    private static final int SEARCH_DELAY_MS = 300;
    private ArrayObjectAdapter mRowsAdapter;
    private Handler mHandler = new Handler();
    private SearchRunnable mDelayedLoad;
    @Override
    public void onCreate(Bundle savedInstanceState) {
        Log.i(TAG, "onCreate");
        super.onCreate(savedInstanceState);
        mRowsAdapter = new ArrayObjectAdapter(new ListRowPresenter());
        setSearchResultProvider(this);
        mDelayedLoad = new SearchRunnable();
    }
```

4. Then, we need to create a `getResultsAdapter()` method that will be responsible for holding the number of rows that have been returned from our search query adapter. This is done as follows:

```
@Override
public ObjectAdapter getResultsAdapter() {
    return mRowsAdapter;
}
```

5. Next, as shown in the following code, we need to create a `onQueryTextChange()` method that will be responsible for refreshing the TV user interface as the user types in a query search string:

```
@Override
public boolean onQueryTextChange(String newQuery) {
    mRowsAdapter.clear();
    if (!TextUtils.isEmpty(newQuery)) {
        mDelayedLoad.setSearchQuery(newQuery);
        mHandler.removeCallbacks(mDelayedLoad);
        mHandler.postDelayed(mDelayedLoad, SEARCH_DELAY_MS);
    }
    return true;
}
```

6. Now, we need to create a onQueryTextSubmit() method that will be called when the user finishes entering his/her search criteria and hits the search button on the TV control pad. This method is created as follows:

```
@Override
public boolean onQueryTextSubmit(String query) {
    mRowsAdapter.clear();
    if (!TextUtils.isEmpty(query)) {
        mDelayedLoad.setSearchQuery(query);
        mHandler.removeCallbacks(mDelayedLoad);
        mHandler.postDelayed(mDelayedLoad, SEARCH_DELAY_MS);
    }
    return true;
}
```

7. Next, we need to create and implement a Runnable class and include a method called setSearchQuery that starts executing the active class. This is done as follows:

```
private class SearchRunnable implements Runnable {
    private String query;
    public void setSearchQuery(String query) {
        this.query = query;
    }
    @Override
    public void run() {
        mRowsAdapter.clear();
        ArrayObjectAdapter adapter = new ArrayObjectAdapter(new
            CardPresenter());
        adapter.addAll(0, MovieList.list);
        HeaderItem header = new HeaderItem(0,
            getResources().getString(R.string.search_results);
            mRowsAdapter.add(new ListRow(header, adapter));
    }
}
```

In the preceding code snippets, we started by adding our import statements that will be responsible for allowing our application to communicate with an Android TV and, just as we did in our CustomRowsFragment class, we incorporated the Leanback support library that provides us with prebuilt components for our TV interface. We then proceeded to extend our TVSearchFragment class using the SearchFragment class, add the code for our onCreate (Bundle savedInstanceState) method that will be called when the fragment has been instantiated, and then set it to handle the returned search results. This is because we have specified that we will be implementing from the SearchFragment.SearchResultProvider class.

In our next step, we proceed to call the `searchRunnable` method that implements the `Runnable` class that includes a method called `run` that starts executing the active class each time the fragment activity is called, and populates our view fragment with the associated row information matching the entered search criteria. This method sets up an `ArrayObjectAdapter` class that instantiates the `CardPresenter` object, adds all movies from our `MovieList` class model, and then displays the movie details that match the entered search criteria.

Creating the custom activity layout resource file

In this section, we will proceed to create our custom layout resource file that will be responsible for ensuring that our custom fragment classes render correctly within the TV user interface:

1. From the **Project Navigator** window, expand the **app** section, select and expand the **res | layout** section.

2. Next, right-click and choose the **New | Layout** resource file menu option as shown in the following screenshot:

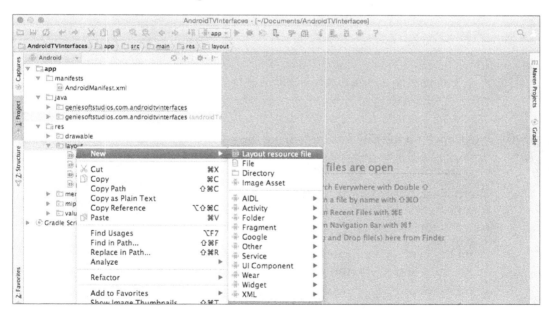

Next, we need to customize the properties of **Resource File** so that it can be used by our application. Here we will need to specify the filename of our layout file and the **Root element** name for our layout information.

3. Enter `activity_custom` for the **File name** field.

4. Next, enter `packageName.CustomFrameLayout` for the **Root element** field as shown in this screenshot:

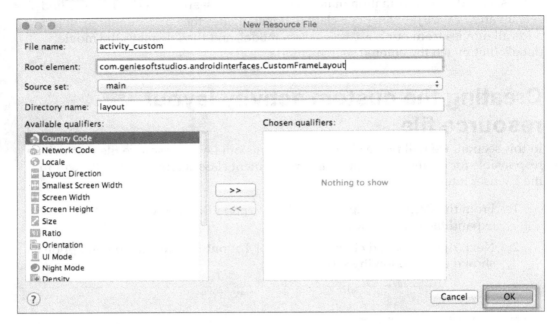

You need to ensure that **Root element** contains the same package name as your project, otherwise you will experience build errors.

5. Next, click on the **OK** button to have the wizard generate the necessary files for you. Once finished, this will open the Android Studio code editor with your custom layout file displayed.

Our next step is to construct the layout for our custom headers and rows fragments, as well as correctly place the search button within our TV user interface.

6. From the **Project Navigator** window, open `activity_custom.xml` that is located in the **res | layout** folder and add the following code:

```
<?xml version="1.0" encoding="utf-8"?>
<com.geniesoftstudios.androidtvinterfaces.CustomFrameLayout
xmlns:android="http://schemas.android.com/apk/res/android"
android:layout_width="match_parent"
android:layout_height="match_parent"
android:clipChildren="false" >
```

```xml
<android.support.v17.leanback.widget.SearchOrbView
        android:id="@+id/custom_search_orb"
        android:layout_width="wrap_content"
        android:layout_height="wrap_content"
        android:layout_marginTop="27dp"
        android:layout_marginLeft="56dp"
        android:layout_gravity="top|left"/>
<FrameLayout
        android:id="@+id/header_container"
        android:layout_width="wrap_content"
        android:layout_height="match_parent"
        android:layout_gravity="top|left"/>
<FrameLayout
        android:id="@+id/rows_container"
        android:layout_width="match_parent"
        android:layout_height="match_parent"
        android:layout_gravity="top|left"
        android:layout_marginLeft="300dp"/>
</geniesoftstudios.com.androidtvinterfaces.CustomFrameLayout>
```

7. Next, from the **Project Navigator** window open `activity_search.xml` that is located in the **res | layout** folder, and add the following code snippet:

```xml
<?xml version="1.0" encoding="utf-8"?>
<fragment xmlns:android="http://schemas.android.com/apk/res/
android"
xmlns:tools="http://schemas.android.com/tools"
android:id="@+id/search_fragment"
android:name="com.geniesoftstudios.androidtvinterfaces.
TVSearchFragment"
android:layout_width="match_parent"
android:layout_height="match_parent"
tools:context=".TVSearchActivity"
tools:deviceIds="tv"
tools:ignore="MergeRootFrame"/>
```

8. Then, from the **Project Navigator** window choose the **app** section, select the `manifests` folder, and then select the `AndroidManifest.xml` file.

9. Now, under the manifest section of the Android TV app, we need to include the permissions to allow our app to run within the handheld device. Enter the following permissions:

```xml
<uses-permission android:name="android.permission.INTERNET"></
uses-permission>
<uses-permission android:name="android.permission.RECORD_AUDIO" />
<uses-feature
    android:name="android.hardware.touchscreen"
    android:required="false" />
```

```
<uses-feature
    android:name="android.software.leanback"
    android:required="true" />
```

10. Next, modify the `<application>` section of the TV app and enter the following highlighted code:

```
<application
    android:allowBackup="true"
    android:icon="@mipmap/ic_launcher"
    android:label="@string/app_name"
    android:theme="@style/Theme.Leanback" >
<activity
        android:name=".MainActivity"
        android:icon="@drawable/app_icon_your_company"
        android:label="@string/app_name"
        android:logo="@drawable/app_icon_your_company"
        android:screenOrientation="landscape" >
<intent-filter>
<action android:name="android.intent.action.MAIN"/>
<category
android:name="android.intent.category.LEANBACK_LAUNCHER"/>
</intent-filter>
</activity>
<activity android:name=".DetailsActivity"/>
<activity android:name=".PlaybackOverlayActivity"
android:theme="@android:style/Theme.NoTitleBar.Fullscreen"/>
<activity android:name=".BrowseErrorActivity" />
<activity android:name=".TVSearchActivity" />
</application>
```

In our next section, we will need to modify the `MainActivity.java` file so that it will be set up to use our `CustomHeadersFragment` and `CustomRowsFragment` files, as well as call our `TVSearchActivity` file when the search icon has been pressed. So let's get started:

1. From the **Project Navigator** window, expand the **app** section, select and expand the **java** section.

2. Next, double-click to open `MainActivity.java` and add the following highlighted code:

```
import android.app.Activity;
import android.app.FragmentManager;
import android.app.FragmentTransaction;
import android.app.UiModeManager;
import android.content.Intent;
import android.content.res.Configuration;
```

```java
import android.os.Bundle;
import android.support.v17.leanback.widget.SearchOrbView;
import android.view.View;
import java.util.LinkedHashMap;
public class MainActivity extends Activity {
    private SearchOrbView orbView;
    private CustomHeadersFragment headersFragment;
    private CustomRowsFragment rowsFragment;
    private final int CATEGORIES_NUMBER = 5;
    private LinkedHashMap<Integer, CustomRowsFragment>
    fragments;
    @Override
    protected void onCreate(Bundle savedInstanceState) {
        super.onCreate(savedInstanceState);
        setContentView(R.layout.activity_custom);
        // Check if we are running on an Android TV Device
        if (isRunningOnTVDevice()) {
            orbView = (SearchOrbView)
                findViewById(R.id.custom_search_orb);
            orbView.setOrbColor(getResources().
                getColor(R.color.search_opaque));
                orbView.bringToFront();
            orbView.setOnOrbClickedListener(new
                View.OnClickListener() {
                @Override
                public void onClick(View view) {
                    Intent intent = new
                    Intent(getApplicationContext(),
                    TVSearchActivity.class);
                    startActivity(intent);
                }
            });
            fragments = new LinkedHashMap<Integer,
                CustomRowsFragment>();
            for (int i = 0; i < CATEGORIES_NUMBER; i++) {
                CustomRowsFragment fragment = new
                    CustomRowsFragment();
                fragments.put(i, fragment);
            }
            headersFragment = new CustomHeadersFragment();
            rowsFragment = fragments.get(0);
            FragmentManager fragmentManager =
                getFragmentManager();
            FragmentTransaction transaction =
                fragmentManager.beginTransaction();
```

```
    transaction
      .replace(R.id.header_container,
      headersFragment, "CustomHeadersFragment")
      .replace(R.id.rows_container, rowsFragment,
      "CustomRowsFragment");
    transaction.commit();
  }
}
// Check if we are using an Android TV Device
private boolean isRunningOnTVDevice() {
  UiModeManager uiModeManager = (UiModeManager)
    getSystemService(UI_MODE_SERVICE);
  if (uiModeManager.getCurrentModeType() ==
    Configuration.UI_MODE_TYPE_TELEVISION){
    return true;
  }
  else {
    return false;
  }
}
public LinkedHashMap<Integer, CustomRowsFragment>
getFragments() {
  return fragments;
}
public void updateCurrentRowsFragment(CustomRowsFragment
  fragment) {
  rowsFragment = fragment;
  }
}
```

In the preceding code snippets, we started by adding our import statements that will be responsible for allowing our application to communicate with Android TV and, just as we did in our CustomRowsFragment class, we incorporate the Leanback support library that provides us with prebuilt components for our TV interface. We then proceed to create our private methods for our searchOrbView, CustomHeadersFragment, and CustomRowsFragment fragment classes, and then modify the onCreate (Bundle savedInstanceState) method to call the setContentView method using our activity_custom layout file when the activity has been instantiated.

In our next step, we call the `isRunningOnTVDevice` method to check to see whether we are running on an Android TV device, then proceed to initialize our search icon properties, and assign an `onClickListener` method so that our `TVSearchActivity` class will be called when this button is clicked. Next, we iterate through the total number of categories that we have specified by the `CATEGORIES_NUMBER` variable while creating an array for each row fragment based on each category.

Then we create a `fragmentManager` object that will be responsible for saving our custom header and row fragment objects, and use the `beginTransaction` method to update the `header_container` and `rows_container` identifiers within our `activity_custom` layout resource file with the name of `CustomHeadersFragment`. We do the same for `CustomRowsFragment` before finally committing the transaction to tell `fragmentManager` that we have finished.

In our next step, we create the `isRunningOnTVDevice` method that will help us determine whether we are running on an Android TV device. We use the `uiModeManager` class and use the configuration type of `UI_MODE_TYPE_ TELEVISION`. We then create a `LinkedHashMap` method that will get each fragment for our header objects and then once the fragment has been updated, it calls the `updateCurrentRowsFragment` method to return the total number of rows contained within each category fragment.

Next, we can finally begin to compile, build, and run our application. Simply press *CMD + F9* and choose your AVD or your Android TV device from the list of available devices as shown in the following screenshot:

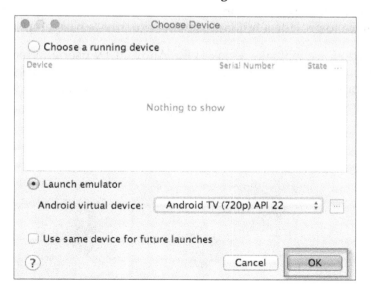

Once the Android TV app has been installed on the AVD, you should see the category information displayed within the side panel along with the associated content information displayed within the middle portion of the TV user interface. Upon clicking the search magnifying glass our custom search activity fragment will be displayed and you can begin searching, as can be seen in the following screenshot:

As you can see, using fragments and classes you are able to customize the Android TV user interface to provide a brilliant viewing experience based on your needs.

In our next section, we will take a look at the Android TV user interface design guidelines and also see what is the importance of designing user interfaces that are intuitive, consistent, and designed with the user in mind to provide the best viewing experience.

The Android TV user interface design guidelines

Designing apps for Android TV need to be different from how you would go about designing apps for phones or tablets, and wearables, as these contain a different user experience. It is important to keep in mind and follow the Android User Interface design principles documentation that Google provides. This document describes the guidelines and principles that help you to design consistent user interfaces and experiences for your Android TV apps, as well as ensure that your application runs efficiently on the Android TV platform, and it involves considering the screen sizes of your custom layouts as well as ease of use your app brings to the platform.

Other areas are covered to ensure the consistency of your application as you navigate from screen to screen, as well as the principles to design good user interfaces. This document also includes some design pattern guidelines that your application needs to conform to. These are as follows:

- You will need to ensure that you design your layouts for the landscape orientation as TV screens always use this viewing orientation. Try to make use of fragment classes to create your user interfaces in sections, and use the GridView class to make better use of the viewing screen space.

- Ensure that you design your artwork assets for best viewing in HD resolution set to 1920 x 1080 pixels and make use of on-screen navigational controls on the left- or right-hand side of the screen.

- TV screen layouts should be simple; it is important to avoid cluttering the user interface by adding sufficient margins and padding between layout controls.

- It is good to provide in-built search functionality within your app as this allows your users to discover new content using Google search features.

- Ensure that your app makes use of the prebuilt fragments that are contained within the v17 Leanback support library classes. These prebuilt fragments provide user interface widgets for TV apps, particularly apps that do media playback and have been specifically designed for use on TV devices with guidance from the Android user experience team.

- When adding typographical text and controls to your Android TV applications, ensure that you use the recommended minimum font size of 12 sp, and the default text size should be set to 18 sp.

There is also information relating to the proper use and appearance of UI elements, such as app and game banners, background images and icons, as well as distributing your app to the Android TV platform.

To obtain further information about these guidelines, it is worthwhile checking out the Android TV design principles documentation at `http://developer.android.com/design/tv/index.html`.

Summary

In this chapter, we learned about Android TV and how we can use this platform to build effective and interactive content by designing and customizing our own interfaces for the Android TV platform.

We got acquainted with and learned more about the `HeadersFragment` and `RowsFragment` classes, and how we can inherit from these classes provided to us by the Android Leanback support library so that we can design our own custom classes for the Android TV platform to provide a nice and clean user experience.

Next, we learned about the `FrameLayout` class and how we use this to extend the capabilities of creating our own layouts so that we can present content professionally within the TV user interface. Then, we looked at how we can incorporate searching capabilities within TV user interface using the `SearchFragment` class that provides us with the callback methods to query search criteria and return content.

We then moved on and learned how to create a custom resource layout file that makes use of our `CustomLayout` class to render our custom header and row information within the Android TV user interface. Finally, we spent some time learning about the design guidelines for Android TV and the design considerations developers need to consider when designing their user interfaces.

This was the final chapter. I hope that you had a ton of fun working through this book, have learned a lot, and have got your Android wearable projects started off on the right foot. Now, you have a wealth of experience with Android Wear and know what it takes to build rich and engaging apps for the Wearable platform using a host of exciting concepts and techniques that are unique to Android's Wear platform.

If, like me, learning about all these cutting-edge technologies and concepts has got you to overflow with ideas, then I can't wait to see what you would build!

Thank you for purchasing this book and I wish you the very best of luck with your Android Wear adventures.

 For any development support questions, you can contact the Android Developer support resources forum at `http://developer.android.com/support.html`.

Index

B

basic notification, for wearables
  blank activity, adding 25, 26
  blank activity, customizing 25, 26
  creating 23
  dependencies, adding to Gradle
      scripts 26-33
  form factors, specifying 24
Bluetooth
  Android wearable app, debugging
      over 69-71
Bluetooth Low Energy (BLE) 4

C

camera
  accessing, through Google Glass 144-148
Confirmations documentation,
     Android Wear
  URL 48
connections
  establishing, for mobile activity 99-101
custom notification
  creating, for wearables 33-39
custom watch face service class
  creating 60-68

D

DataApi
  used, for receiving image
      data 114-119
design guidelines, Android TV user
     interface
  about 193
  URL 194

G

Google Glass
  app, launching within 157
  camera, accessing through 144-148
  Google Maps API, incorporating
      with 149-154
  voice input, incorporating within 141-144

Google Glass application
  AndroidManifests file, configuring 137-139
  creating 127-130
  custom camera layout resource file,
      creating 139-141
  custom menu resource file,
      creating 134-137
  theme, setting for 131
Google Glass Development Kit (GDK)
  about 123, 157
  functionality, for Glass 158
  URL 158
Google Glass Development Kit SDK
  installing 122-124
Google Glass main activity UI
  modifying 154-157
Google Glassware principle design
     guidelines
  about 160, 161
  URL, for documentation 161
Google Maps API
  incorporating, with Google Glass 149-154
Google Mirror API 157
Google USB Driver
  installing, for Windows 124-127
  URL 127
Gradle 27

I

image data
  receiving, DataApi used 114-119
  transferring, to Android wearable 111-113
immersion activity 130
information
  presenting, inside WatchFace class 55-60
installing
  Android Wear support library 7
  Google Glass Development
     Kit SDK 122-124
  Google USB drivers, for Windows 124-127

J

Java Runtime Environment (JRE) 6

Thank you for buying
Android Wearable Programming

About Packt Publishing

Packt, pronounced 'packed', published its first book, *Mastering phpMyAdmin for Effective MySQL Management*, in April 2004, and subsequently continued to specialize in publishing highly focused books on specific technologies and solutions.

Our books and publications share the experiences of your fellow IT professionals in adapting and customizing today's systems, applications, and frameworks. Our solution-based books give you the knowledge and power to customize the software and technologies you're using to get the job done. Packt books are more specific and less general than the IT books you have seen in the past. Our unique business model allows us to bring you more focused information, giving you more of what you need to know, and less of what you don't.

Packt is a modern yet unique publishing company that focuses on producing quality, cutting-edge books for communities of developers, administrators, and newbies alike. For more information, please visit our website at www.packtpub.com.

About Packt Open Source

In 2010, Packt launched two new brands, Packt Open Source and Packt Enterprise, in order to continue its focus on specialization. This book is part of the Packt Open Source brand, home to books published on software built around open source licenses, and offering information to anybody from advanced developers to budding web designers. The Open Source brand also runs Packt's Open Source Royalty Scheme, by which Packt gives a royalty to each open source project about whose software a book is sold.

Writing for Packt

We welcome all inquiries from people who are interested in authoring. Book proposals should be sent to author@packtpub.com. If your book idea is still at an early stage and you would like to discuss it first before writing a formal book proposal, then please contact us; one of our commissioning editors will get in touch with you.

We're not just looking for published authors; if you have strong technical skills but no writing experience, our experienced editors can help you develop a writing career, or simply get some additional reward for your expertise.

Learning Android Forensics

ISBN: 978-1-78217-457-8   Paperback: 322 pages

A hands-on guide to Android forensics, from setting up the forensic workstation to analyzing key forensic artifacts

1. A professional, step-by-step approach to forensic analysis complete with key strategies and techniques.

2. Analyze the most popular Android applications using free and open source tools.

3. Learn forensically-sound core data extraction and recovery techniques.

Android NDK: Beginner's Guide
Second Edition

ISBN: 978-1-78398-964-5   Paperback: 494 pages

Discover the native side of Android and inject the power of C/C++ in your applications

1. Create high performance mobile applications with C/C++ and integrate with Java.

2. Exploit advanced Android features such as graphics, sound, input, and sensing.

3. Port and reuse your own or third-party libraries from the prolific C/C++ ecosystem.

Please check **www.PacktPub.com** for information on our titles

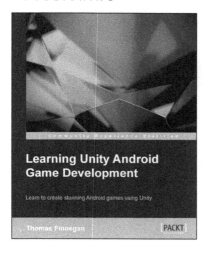

Learning Unity Android
Game Development

ISBN: 978-1-78439-469-1   Paperback: 338 pages

Learn to create stunning Android games using Unity

1. Leverage the new features of Unity 5 for the Android mobile market with hands-on projects and real-world examples.

2. Create comprehensive and robust games using various customizations and additions available in Unity such as camera, lighting, and sound effects.

3. Precise instructions to use Unity to create an Android-based mobile game.

Learning Android
Application Testing

ISBN: 978-1-78439-533-9   Paperback: 274 pages

Improve your Android applications through intensive testing and debugging

1. Focus on Android instrumentation testing to ensure full application coverage.

2. Apply testing techniques and utilize tools to improve Android application development.

3. Build intensively tested and bug free Android applications.

Please check **www.PacktPub.com** for information on our titles